Earthy Realism

The Meaning of Gaia

Edited by

Mary Midgley

SOCIETAS
essays in political
& cultural criticism

imprint-academic.com

Published in the UK by Societas
Imprint Academic, PO Box 200, Exeter EX5 5YX, UK

Published in the USA by Societas
Imprint Academic, Philosophy Documentation Center
PO Box 7147, Charlottesville, VA 22906-7147, USA

ISBN 9781845400804

A CIP catalogue record for this book is available from the
British Library and US Library of Congress

Contents

Contributors

Dr Richard Betts leads the Climate Impact Research Group in the Hadley Centre for Climate Change at the Met Office. He has a degree in Physics, a Masters in Meteorology and Applied Climatology, and a PhD in Meteorology for his thesis on 'Modelling the Role of the Vegetated Land Surface in Climate and Climate Change.' He has written or co-authored fifty books and articles. He was a lead author on the Millennium Ecosystem Assessment and the fourth Assessment Report of the Intergovernmental Panel on Climate Change.

Elaine Brook is a writer and photographer who spent 12 years living and working in remote areas of the Himalaya, where her knowledge of both Nepali and Tibetan enabled her to communicate easily with the people she met. She currently teaches meditation and deep ecology and gives talks and workshops bringing together insights from the arts and science. She is Director of the Gaia Partnership, a not-for-profit organization dedicated to education and awareness-raising on ecological issues. Her books include *The Windhorse, Land of the Snow Lion*, and *In Search of Shambhala*, published by Jonathan Cape. More details; www.gaiapartnership.org.

Susan Canney has worked in science, conservation and environmental policy for over 20 years. She has MAs in Natural Sciences, Landscape Design and Environmental Policy, and a doctorate in Conservation Ecology. Her current work is in using spatial tools such as remote sensing and geographical information systems to devise conservation strategies and guide their implementation. She teaches Global Ecology at the University of Oxford; is Secretary of the Gaia Network and of the Earth Systems Science Special Interest Group at the Geological Society of London, and is a part of Forum for the Future's 'Reconnections' team.

Cormac Cullinan is a Cape Town based environmental attorney and author of *Wild Law* (Green Books, 2003). He is a graduate of the University of Natal and of Kings College, London and is an honorary research associate of the University of Cape Town. He has lectured and written widely on governance issues relating to human interactions with the environment. Cormac has worked on environmental governance related projects in more than 20 countries and has particular expertise in drafting legislation and designing governance systems. He currently manages a consultancy that advises on environmental governance issues (see www.enact-international. com) and is a director of a leading South African environmental law firm (seewww.winstanleycullinan.co.za).

Maggie Gee is the author of ten novels including *The Burning Book, Light Years, Grace, Where Are The Snows?, The Ice People, The Flood* and *My Cleaner*. Her most recent publication is a collection of short stories, *The Blue* (Telegram 2006). *The White Family* was short-listed for the Orange Prize and the International Impac Prize. Her work has been translated into 11 languages and she is the first female Chair of the Royal Society of Literature. She is Visiting Professor of Creative Writing at Sheffield Hallam University.

Brian Goodwin was born in Canada where he studied biology. He then took a mathematics degree at Oxford and a Ph. D. involving biology and mathematics at Edinburgh University. He has held research and teaching positions at MIT, at the University of Sussex, and the Open University, UK, where he was Professor of Biology. He was connected with the Santa Fé Institute for a number of years in the 80s and 90s. He now teaches Holistic Science at Schumacher College in England. His interests are in developing a science of qualities that can address issues of health and quality of life in diverse areas, in promoting holistic patterns of living, and in the reunion of the arts and humanities with the sciences. His best-known book is *How The Leopard Changed Its Spots: The Evolution Of Complexity* (Phoenix 1995).

Dr Stephan Harding holds a doctorate in ecology from the University of Oxford. He taught wildlife ecology at the National University in Costa Rica and is now the Resident Ecologist at Schumacher College, where he is also the co-ordinator of the MSc in Holistic Science. In his spare time, Stephan teaches ecology all over the world, and has recently been appointed co-holder with James Lovelock of the Arne Naess Chair in Global Justice and the Environment at the University of Oslo, Norway. Stephan is the author of *Animate Earth, Science, Intuition and Gaia* (Green Books, 2006)

John Mead, after war service with the Royal Engineers, studied Philosophy, Politics and Economics at Oxford, and Social Psychology at LSE. He worked for five years for the National Coal Board and thus met in 1956 E.F. Schumacher (author of *Small is Beautiful*, and founder of the Intermediate Technology Development Group) by whom he was much influenced. In due course he became a Member of ITDG (now Practical Action). After a number of posts in education, he lectured in social psychology at what is now London South Bank University. He later studied with the British Association of Psychotherapists and practiced as a Jungian psychotherapist. He has a special interest in the psychological aspects of the ecological crisis, in particular climate change. He was a member of a working party formed to advise the bishops of London and Southwark on environmental policy in their dioceses. He is the author of a number of articles and papers on environmental matters and has given presentations to a range of audiences.

David Midgley studied philosophy at Manchester, Oxford and York. From 1996 to 2001 he was Director of Jamyang Buddhist Centre Leeds, where he continues to teach Buddhist philosophy and meditation. With Mary Midgley and Tom Wakeford, he founded the Gaia Network, an interdisciplinary group dedicated to exploring the intellectual and social implications of Gaia Theory. He now works at Leeds Development Education Centre, is engaged in research into philosophical aspects of Gaia Theory, and is active in the movement for local sustainable development.

Mary Midgley (DLitt) is a philosopher whose special interests are in the relations between humans and the rest of nature (particularly in the situation of animals), in the sources of morality, and in the relations between science and religion (particularly in cases where science becomes a religion). In recent years she has become much interested in the concept of Gaia, in which all these concerns come together. Until her retirement she was a Senior Lecturer in Philosophy at the University of Newcastle on Tyne in England, where she still lives. Her husband, Geoffrey Midgley, also a philosopher, died in 1997. She has three sons. Her most recent books are *Science and Poetry* (2001), *The Myths We Live By* (2003), and a memoir, *The Owl of Minerva* (2005), all from Routledge, London.

Anne Primavesi, formerly Lecturer and Research Fellow in Environmental Theology at the Department of Theology and Religious Studies, University of Bristol, and Research Fellow at the Centre for the Interdisciplinary Study of Religion, University of London, is currently Fellow of the Westar Institute, California and a Founding Research Fellow of the Lokahi Foundation, London. Her published books include *Sacred Gaia; Holistic Theology and Earth System Science* (Routledge 2000), *Gaia's Gift; Earth, Ourselves and God After Copernicus (Routledge 2003)*, and *Making God Laugh; Human Arrogance and Ecological Humility* (Polebridge Press 2004). She contributed the entry on 'Gaia' in *Encyclopaedia of Religion* (Macmillan Reference, 2nd ed., 2004), the entry on 'Ecofeminist Theology' in *Cambridge Dictionary of Christianity* (Cambridge University Press 2006) and the entry on 'Ecology' in the *Blackwell Companion to the Bible and Culture* (2006).

Patricia Spallone worked for 12 years as a biochemist at the University of Pennsylvania Medical School (US) before moving to Britain where she turned her attention to women's health and reproductive rights, and the social studies of science and technology. She worked most recently in the Wellcome Trust's Biomedical Ethics Program, and then as Associate Director of the BIOS Centre, London School of Economics. Now working independently from her home in Norwich, she is Honorary Visiting Fellow in the Centre for Women's Studies, University of York.

John Turnbull was formerly the Network Associate (co-ordinator) for the Worldwide Democracy Network (WDN), the central hub of a network of people interested in issues of systems-thinking, politics and economics. He has a degree in economics from the University of Newcastle on Tyne and currently works in the UK civil service.

David M. Wilkinson is Reader in Environmental Science in the School of Biological and Earth Sciences at Liverpool John Moores University. He has wide research interests, which include both theoretical and empirical work. Much of his theoretical work is on the interface between ecology and evolutionary theory; this has included studies of the evolution of mutualisms (mutually beneficial interactions between organisms). Thus he has written a series of research papers on the evolution of Gaia; the biggest putative mutualistic system. Other interests include biogeography, microbial ecology, the history of science (mainly Biology and Geology) and environmental archaeology. His recent book on *Fundamental*

Processes in Ecology (Oxford University Press 2006) attempts to recast ecological theory in a Gaian context, in order to develop a conceptual approach to ecology that is more suited to addressing the major problems likely to be produced by climate change.

John Ziman (FRS) was born in New Zealand in 1925. He studied at Oxford and lectured at Cambridge before becoming Professor of Theoretical Physics at Bristol in 1964. In 1967 he was elected to the Royal Society for his research on the electrical properties of metals. After taking voluntary early retirement from Bristol in 1982 he gave profound attention to investigating the relation between science, technology and society and wrote a number of ground-breaking books on that subject. For many years he was chairman of the Council for Science and Society and between 1986 and 1991 he headed the Science Policy Support Group. He died in 2005.

James Lovelock

Foreword

Looking back into Gaia's long history we find the occasional emergence of species that wreak havoc and benefit on a planetary scale. The photosynthesizers who first split water to release oxygen were among the first. Oxygen is carcinogenic, but without it the Earth would now be a dry desert; the oxygen in the air prevents the hydrogen of water from escaping into space as it has done on Mars and Venus. Then (I would speculate) there may have been a first marine organism that countered the harmful saltiness of the ocean by synthesizing the charge-neutral salt dimethyl sulphonio propionate (DMSP). When these individuals died, the DMSP was released into the ocean where it decomposed to form dimethyl sulphide (DMS), a compound volatile enough to escape from the ocean into the air. Their success against salt may have been short-lived because, as they bloomed across the world oceans, a huge flux of DMS may have so clouded the Earth that it cooled almost to a snow ball and reduced all life to a mere fraction. In time, consumers of DMSP and DMS appeared and Gaia struck a balance, now using DMS as part of the beneficial air-conditioning of the Earth. Could our own disastrous destruction of ecosystems and pollutions be merely the beginning of another learning phase of the Earth, one in which Gaia learns how to deal with and benefit from a collective intelligence?

To better understand our current crisis of global heating, it may be helpful to take this long-in-time-and-space view and see ourselves and the Earth from afar. Not least, it allows us to escape from the narrow confining scientific tendency to reduction. Most important, it sets us free to wonder if the Earth is alive. Scientifically correct biologists define life as 'something that reproduces and corrects the errors of reproduction through natural selection' but that must be far too limiting a definition. At the lowest level in Physics it could be said that anything that has a lifetime must be alive; so is the atom of a radioactive element alive? Is mortality or impermanence part of the definition of life? Physicists have defined life as an open but

bounded system that sustains low internal entropy. By this definition, Gaia, the system of all life and the Earth's surface environment,
is alive. Much more than that, Gaia has all the usual properties of life;
it metabolises, it self-regulates its climate and chemical composition,
and has sustained for nearly 4 billion years an utterly improbable
low entropy. Does something that has lived a third of the age of the
universe need to reproduce? Wouldn't it be an embarrassment if it
did? Mars and Venus started out with chemical compositions similar to the Earth but look at them now. Both are unspeakably inhospi-
table deserts and so would the Earth have been had not life emerged.
Sadly, the Earth will die and become like its dead siblings in about
500 million years — or sooner if we continue as we are doing now.

Perhaps the single most important thing that we can do to undo
the harm we have done is to fix firmly in our minds the thought: *the
Earth is alive*. Once such a thought becomes instinctive we would
know that we cannot cut down the forests for farmland to feed ourselves without risking the destruction of our home planet. Farmland
and tree plantations cannot serve as a replacement for natural forests
that have evolved with their environment over millions of years and
once served to keep the climate tolerable and the air good to breathe.

Mary Midgley

Introduction
The Not-So-Simple Earth

Double Vision

The notion of GAIA—of Earth and the life on it as an active, self-maintaining whole—has long given Western intellectuals a terrible squint. One eye has seen this notion as a piece of Science, at first as a mistaken one, then, later, as one that may be partly accepted. But the other eye sees it as something visionary or spiritual, alien to science, perhaps hostile to it—perhaps a new truth, perhaps just a childish 'new age' fancy.

If we want to understand what is now going wrong with our planet we will probably need to bring these two different angles together, rather than just glimpsing the concept confusedly in both ways. This book has grown out of discussions held by a small group called the Gaia Network composed of people who want to start on that work. The work is difficult because the squint, like others of its kind, marks an inner conflict. It is not just a debate between 'two cultures' outside us. It concerns the whole relation between our inner and outer lives. This has long been troublesome because, for much of the twentieth century, the behaviourist approach ordered us to ignore the inner life entirely for scientific purposes, though we still had to take it seriously for everyday affairs.

Today, real efforts are being made to bring these two parts of our lives together. For one thing, during the 1970s, the word 'consciousness' was once more allowed to be freely pronounced in the social sciences. An eager pursuit of 'consciousness studies' followed as psychologists reopened problems that had been closed to them for some fifty years. Neurologists, who had, meanwhile, been discovering remarkable things about the brain, have joined them there and these enquiries now flourish.

Another sign of reconciliation has been the quiet return of the word 'spiritual' to decent intellectual usage during the last two decades. For reasons that are still not quite clear, even card-carrying atheists seem to have noticed that this word describes a crucial area of our lives, an area reaching far beyond the objectionable parts of traditional religion. It is the province of our responsive and creative imagination — not just a fiction-factory but a vitally necessary place where we work out the interpretative patterns we need for our life-world as a whole, structures and visions to provide some usable order in the chaotic world of our experience.

This is where the concept of Gaia belongs. This is why it has both a scientific and a spiritual aspect. It is one of the new, constructive visions that we shall need as we work to replace the dramatic New-tonian world-picture that was, for so long, the background of our thought. That picture always centred on the powerful image of *clock-work*, leading us to look for what (significantly enough) we still call 'mechanisms' — measurable causal sequences, preferably involving interactions by pushing and pulling between solid, inert particles. We know now that we need to find other kinds of pattern besides these if we are to give sense and meaning to our lives. That need is already plain within science itself because modern physics no longer uses the old mechanistic notion of matter at all. It is not now 'reductive' in the atomistic sense of looking for the final truth in ulti-mate solid particles. Instead, it deals in various kinds of forces and fields and in webs of connection, rather than in separate items to be connected.

In Paradigm Country

Since physics has been the scene of this deep change, it is not surpris-ing that it is also the place where the importance of these shifting background visions, or paradigms, was first noticed. Forty years ago Thomas Kuhn pointed it out in *The Structure of Scientific Revolutions*, and its crucial importance for changes in science is now accepted. But its significance is far wider than this scientific context. The point is not just that one background framework in physics gives way to another but that, whenever we deal with complex phenomena, we may need to use several such frameworks together.

Pluralism pays. We get on better by combining a number of differ-ent ways of thinking than by signing up for just one of them. Thus, the Newtonian vision with its attendant imagery was not forgotten when the Einsteinian one came in, nor did the still older, everyday ways in which we think about physical objects become unnecessary. They all go on being used side by side, each being appropriate in its own sphere. Similarly, we will get on much better by combining the various ways in which we can use the concept of Gaia to perceive the earth, and grasping the connection between them, than we will by treating them as alternatives.

This is urgent business. The way in which we think of the earth is not marginal; it deeply affects our conception of ourselves. Since the Industrial Revolution, our culture has largely treated the physical world as an inert heap of resources — a convenient store provided to power our various projects. Accordingly we have thought of ourselves as pure independent intellects, active beings detached from the rest of nature and licensed to use it as we please.

Of course that bold approach produced some splendid achievements. But now, as we start to see the damage we have unconsciously been doing to the world, our whole image of ourselves is been thrown into confusion. We don't see how to make sense of this new situation, so we tend not quite to believe in it.

This sense of unreality almost inevitably accompanies paradigm-changes — points when we realise that our world view is inadequate but don't yet see what to do about it. In these emergencies, the first thing needed is often to become more aware of our existing world-view so as to grasp what has gone wrong with it.

Mechanism Lives

Today, this means understanding better the confident, dominant, mechanistic attitude to the physical world which has brought us into this trouble. Though we call that attitude 'modern', it actually dates from the seventeenth-century scientific revolution. That is when we began to see science both as our main tool for dominating nature and as our prime glory, the central justification for our empire. It is interesting that what happened at this point was not just a general exaltation of Science itself but a victory of one scientific faction — the mechanists — over another, equally reputable at the time. That other tradition preferred the imagery of *sympathy* and *antipathy* between various substances and often used the language of *love*, a universal network of bonds generated by Mother Nature. Thus Marsilio Ficino:

> The parts of this world, like the member of one animal ... are united among themselves in the community of a single nature .... From their communal relationship a communal love is born and from this love a common attraction; and this is the true magic

... . Thus the loadstone attracts iron, amber, straw; brimstone,
fire: the Sun draws flowers and leaves towards itself, the moon,
the seas ... . (Ficino, 1956, p. 220)

This was not just magical talk. Scientists used this approach to compare and understand various kinds of attraction and they could
surely have gone on using it to explore the phenomena of electricity.
Kepler, who liked this method, eagerly suggested that the moon's
attraction might be responsible for the tides. But the mechanists,
including Galileo, rejected that as impossible. And Newton, when
he did, later, endorse gravitation, ruled that it must be due to God's
intervention, since it was not possible for one object to attract
another directly.

This mechanist ideology is not just a historical curiosity. It sur-
vives, underlying the alienation and contempt towards the physical
world which many still take for granted today. It centred on regard-
ing matter as inert — dead — and thus something to which we ought
to be indifferent. It was also particularly hostile to the idea of Nature
as a nurturing mother. Thus Descartes, in expounding the mechanistic paradigm, insisted that form and movement in the physical
world are not due to any productive force in it but are the direct gift
of spirit — that is, of God. There is no point at which 'nature' could
intervene. 'Know', he wrote, 'that by nature I do not mean some goddess or some other sort of imaginary power. I employ this word to
signify matter itself' — that is, inert stuff. Those who have understood his doctrines will (he says) 'in future, *see nothing ... whose cause*
they cannot easily understand, nor anything which gives them any reason
to marvel' (Descartes 1973, p. 349). Magnetism will eventually be
explained merely in terms of the 'size, shape and motion of different
particles of matter'. And any apparent mysteries about life itself will
soon be resolved, as it becomes clear that animals are really only simple, unconscious automata.

Thus, the mechanistic paradigm confirmed the Christian teaching
that any reverence for nature was mere idle superstition. Science
joined religion in declaring that the natural world did not deserve
such exaltation. Life, in particular, was not mysterious but an ordinary slight disturbance produced by simple machinery which
would soon be fully explained. Students of biology need not, therefore, use any distinctive way of thinking. The methods of physics
and chemistry were all that was required for scientific explanation.

This disillusioned view of matter might have been expected to
make mechanistic campaigners like Bacon stop personifying Nature

altogether. Far from that, however, they went on for a long time enthusiastically treating her as female and suggesting new ways of attacking her, searching out her inmost secrets, piercing her armour and generally bringing her into submission. The first manifesto of the Royal Society proclaimed the founding of 'a truly masculine philosophy'. And this imagery of having won the gender-battle long continued to accompany scientistic celebrations of human dominion over the earth around us.

Things don't, of course, look quite like that today. Like a hermit crab growing out of its shell, Western culture now feels uncomfortable about a whole set of anomalies which keep contradicting its familiar paradigm. That unease doesn't, of course, automatically produce a new one. But the situation is unstable. If James Lovelock and Lynn Margulis had not put forward the notion of Gaia, then somebody else would surely have had to do something else of the kind.

What Comes Next?

1. *The Science*

What was needed was to feel once again that we are actually part of the earth's system, as dependent on it as on the air and temperature around us. Though we can disturb that system, we can neither control it nor exist without it. (Thus, both the idea of *spaceship earth* and the idea that we are earth's *stewards* are too grandiose, though of course they are both better than thinking that we are its asset-stripping owners). In order to take in this unwelcome fact, we had to become aware that this system existed in the first place. But, as Lovelock has pointed out, it was extraordinarily hard for modern science to grasp this large fact because it was sub-divided into a crowd of non-communicating tribes, none of whom saw any particular need for the exercise.

In that climate, it is not really surprising that it took scientists about thirty years to accept the basic idea of the Gaian system. Today, however, they have done so. The core scientific point of the story—the cycle by which living things deeply and constantly affect the inorganic world rather than just being its helpless passengers, so that the two form a continuous whole—is now widely accepted. It finds expression in the numerous departments of Earth Science which have grown up in our universities, bringing biology and geology together to study the continuity of the cycle.

There are, of course, still controversies about just how far life is responsible, through these effects, for maintaining the conditions that make possible its own survival. But the central surprising fact, to which Lovelock originally drew attention, remains. It is that these quite complicated conditions have, so far, actually been continuously maintained, in the face of many adverse events, ever since life first appeared on earth. The planet as a whole has managed to keep its climate and temperature within the narrow range needed for life — despite changes such as the sun's becoming hotter by more than a quarter — over three-and-a-half eons, by contrast with Mars and Venus where the temperature and atmosphere quickly became irretrievably hostile. This remains extremely impressive.

2. *The Imagery; Personification and Dependence*

What, however, about the imaginative aspect? Was it necessary to use language that apparently personified the earth?

It seems that it really was so. Before Lovelock used the name Gaia, people could not really grasp his ideas at all. His friend, the novelist William Golding, suggested the name of the ancient Greek mother-goddess, Gaia. And when Lovelock tried that name, people began to understand him. What personification does is to attack the central, disastrous feature of the mechanistic paradigm which we have just noticed — the conviction that the physical world is inert and lifeless. It reverses the propaganda which had dramatized that idea by using machine-imagery and by fiercely denouncing alternative, more personal images such as Mother Nature.

Today, we should be quicker than seventeenth-century people were to see through that propaganda. We know now that science always uses imaginative visions or paradigms which change from time to time, along with the imagery that expresses them. What matters is to grasp what the changing images mean.

Personifying the earth means that it is not just a miscellaneous heap of resources but a self-maintaining system which acts as a whole. It can therefore be *injured*; it is *vulnerable*, capable of health or sickness. And, since we are totally dependent on it, we are vulnerable too. Our deep, confident, seventeenth century conviction — expressed in a lot of space-literature — that we are really independent minds, essentially detached from a planet which we can easily exchange at any time for another one, has been a fantasy. Like babies, we are tiny, vulnerable, dependent organisms, owing our lives to a tremendous whole. That is surely what the Greeks meant to acknowledge. and what our own ancestors meant when they spoke of Mother Nature.

This idea is really surprising today because the whole propaganda of the Enlightenment has urged us to be *adult* — to grow up, to reject our parents' standards, to assert our independence against every kind of authority. Even the fatherhood of God, which the original mechanists readily accepted, is increasingly seen as something of an intrusion. And in social life this campaign for independence has, of course, often been useful. But, applied to the natural world, it is surely a trifle crazy. In this book, we have tried to suggest what the world might look like without it.

Sources

Descartes, René (1973), 'Le Monde'. In F.Alquié (ed.), *Oeuvres Philosophiques de Descartes*, vol. 1 (Garnier Frères).
Ficino, Marsilio. *Commentaire sur le Banquet de Platon*, tr. R.Marcel (Paris).

John Ziman[1]

The Challenging, Inspiring, Irreducible Pluralism of Gaia

One of the disconcerting things about Gaia is that scientists find it so very disconcerting. Is this because it can't be squeezed into any of their established pigeonholes? It mixes together concepts from the chemical, biological and physical sciences. Why is it so difficult to combine these into a coherent, unified representation or vision?

I argue that this intrinsic pluralism is one of its glories and fascinations. Think historically. The planet Earth assembled, imbricated and remodelled itself by purely physico-chemical processes. For a billion years or so, everything that happened could be described in the language of gravitational forces, thermodynamic phases, chemical compounds, etc.

Then, life emerged. Novel entities, with unprecedented properties — i.e. distinct *organisms* — appeared on the scene. To describe their phenomenology required a whole new conceptual vocabulary. Thus, the further history of Gaia had to be written, in part, in the language of biology. This had to include a great many absolutely basic terms such as organism, function, behaviour, metabolism, perception, predator, survival, epigenesis, descendant, evolution, etc. for which there were no equivalents in the language of the physical sciences.

In due course, a million or so years ago, another conceptual fulguration occurred. The emergence of *consciousness* enabled hominids to

---

[1]   John Ziman conducted discussion on this brief piece (written in April 2004) at the Gaia Network shortly before his last illness. If he had lived he would surely have expanded it, but, just as it stands, it still seems suitable to stand as a prologue to this book. MM

engage in another completely unprecedented phenomenology. This, again, is strongly influencing the career of Gaia. So yet another new language is required, with terms for *social* concepts such as symbol, communication, idea, institution, plan, story, contract, money, leader, love, death, kin, ancestor, god, action, etc. These, in turn, have no strictly biological counterparts—although, of course, they are also subject to the necessities of their physical and biological dimensions.

So now we have to make sense of a world containing entities of these three different kinds, each governed by a different 'logic' and defined in a different conceptual language. These differences have arisen naturally, are universally recognised in natural languages, and have given rise to different bodies of scientific knowledge. And, because the successive events of their emergence were entirely unpredictable, as was what emerged at each stage, these phenomenologies, logics, languages and sciences are irreducibly distinct, and cannot be unified into a single formal system. The pluralism of the sciences is not just a weakness of the human intellect: it is a product of the physico-bio-psychic history of our Gaian abode.

Many of the most interesting—and deeply puzzling—topics in modern science relate to the 'boundary objects' that straddle these domains, such as genes, artefacts, information, mind/brains, etc. Of all these, Gaia is the most intriguing, not only as the birth-mother of all three domains, but as a natural entity centred on a 'triple point' where they all meet. Thus, questions such as 'Is Gaia alive?', 'Did Gaia evolve?' or 'Can Gaia be saved?' are so constrained terminologically that they cannot be answered meaningfully. Nevertheless they challenge our capacity to make interpretative connections across the frontiers of incommensurability between the plurality of domains of scientific knowledge.

Brian Goodwin

Gaia and Holistic Science

The Gaia Hypothesis belongs to a twentieth-century dialectic within the sciences that has been called by different names such as nonlinear dynamics, systems theory, and Complexity Theory, but I shall refer to this transforming context as Holistic Science, a coming together of different disciplines and different ways of knowing into a new unity. It has been emerging throughout much of the century and continues to exert a profound influence on the way in which science is understood and is practised. It continues to extend the boundaries of scientific inquiry in ways that are more and more inclusive, so that particular ways of defining scientific inquiry cease to capture the essence of systematic procedures of knowing and doing in our culture. This essay explores these developments, particularly as they relate to the rapidly-changing challenges facing humans of living in sustainable and harmonious relationship with nature.

James Lovelock's Gaia Hypothesis emerged in the 1960s at the same time as another concept that was destined to transform science in quite unexpected ways. This was the idea of chaos in dynamical systems. Although dynamical chaos was born in the nineteenth century in the mind of the great French polymath, Henri Poincaré, it didn't surface into general scientific awareness until Edward Lorenz carried out his simulations of simplified equations describing weather patterns at MIT in the 1960s, revealing the famous phenomenon of the butterfly effect, or sensitivity to initial conditions. This exorcised Laplace's demon from science, the spectre of a mechanical and predictable cosmos that has haunted cosmology since Descartes. The planets themselves have chaotic aspects to their dynamics, resulting in unpredictable behaviour. The effect of Gaia theory was equally dramatic: it transformed our way of thinking about evolution. It is interesting to note that Chaos and Gaia emerged into general consciousness at about the same time, a period that is also memorable for other events that I shall bring into the picture later.

The Gaia Hypothesis was independent of chaos, and in fact Lovelock sees no significant role for this type of unpredictability in his dynamical picture of the earth as a continuously evolving unity. His holistic concept belongs rather within a framework that emerged with clarity and precision only in the 1980s: that of Complexity Theory, championed and effectively promoted by the Santa Fé Institute and taken up by a remarkable range of interested parties, from business organisations to the humanities. Gaia belongs within Complexity Theory, which is characterised by emergent properties that are not predictable in advance from complex dynamics, requiring computer simulation to be demonstrated. These emergent properties are then perfectly repeatable providing the composition and interactions of the system are not altered. However, any small change in the components of the system or in their pattern of interactions can result in a change in behaviour that is not predictable in advance. This is of course the basis of the models of climate change that have proliferated, gaining greater and greater complexity as more variables and interactions are added to make the models more realistic, with the discovery that there are patterns of change unfolding whose consequences no-one can foresee. It is possible only to forecast general trends, such as progressive Arctic warming, not detailed events such as exactly when the Arctic ice sheet will disintegrate irreversibly. The success of Gaia in the form of Earth System Science is now generally acknowledged. We have lived through a transformation in the way we think about planet earth that is remarkable, and highly significant for the way in which we now use our scientific knowledge. In this essay I shall examine the place of Gaia Theory in an extended form of science that we call Holistic Science at Schumacher College. This embraces ways of knowing that are subject to the same type of consensual evaluation that are used in conventional science, but includes qualities and values as reliable indicators of the condition of complex wholes. This is in accord with general principles of common sense. There are other consequences of this approach that alter our perception of evolution in ways that go well beyond the insights of Gaia theory, as will be explored.

Gaia and Evolution

The radical step that Lovelock took that landed him in deep trouble with biologists was to declare that life has a significant impact on the state of the earth as a dynamic system, affecting variables such as atmospheric oxygen levels, temperature, and salt concentrations in the oceans. The conventional Darwinian view is of course that geological and chemical processes determine the basic properties of the planet and life either adapts to these or perishes. Not so, claimed Lovelock. The activities of organisms on the planet, especially the microbes (as pointed out by Lynn Margulis, Lovelock's initial collaborator), alter global properties, and they accumulated evidence that soon became totally convincing. Comparing the condition of Planet Earth with that of other planets made it perfectly clear that it is the presence of life that causes the exceptional dynamic properties that keep our planet in a state compatible with life. There is no spooky prescience or foresight that life exercises in holding temperature or water at levels that allow life to continue; it is just the dynamics of a complex system with intricate feedback loops, as Lovelock showed through an ingenious model that demonstrated how temperature regulation can come about through perfectly natural means. However, the values of the global variables that are crucial for life can and have changed in ways that challenge existing life. This happened when there was a sudden increase in oxygen levels after bacteria discovered how to carry out photosynthesis. Instead of the reducing atmosphere to which bacteria were accustomed, oxygen levels rose dramatically, presenting a major challenge to microbial metabolism because of the dangerous free radicals that oxygen produces in cells. Microbes had to change their ways of releasing energy from molecules, making use of oxygen as a electron acceptor. The challenge was met, and life went on to glorious new patterns of order in the form of an immense diversity of complex species. Life has always faced such difficulties, but the resources have so far always been there to find a solution.

We now face a challenge that is going to test our culture to the limit, and perhaps beyond. By forcing up global temperatures through the release of carbon dioxide from previously buried fossil fuel, we are pushing the earth to a transition point that will probably be uncomfortably hot and dry for us. The Earth has faced such a situation before, but it took half a million years or so before it cooled down again to global temperatures in a more hospitable range, from our point of view. There is no guarantee that it could do so again, so we are entering new territory from which we have no exit strategy. This can be read as either a terminal disaster or as an opportunity to adopt a much more sensible and sustainable life-style. The Industrial Economic Culture that we have embraced is one that bears within itself the seeds of its own destruction, as we can see very clearly through the consequences of our profligate way of using natural resources. These are present in abundance, though we have managed to focus on a series of toxic and damaging technologies that are destroying natural abundance through greedy harvesting techniques and pollution of natural systems. Once these dangerous technologies are reduced and replaced, natural cycles can be restored and the earth can regenerate a healthy condition.

However, this requires a dramatic change of focus on our part. We have the technological know-how to achieve this, but do we have the will to go through the transition before there is a sudden catastrophic collapse of the economic and social systems on which our culture depends? I personally believe we have, and that there is already a critical mass of people with the knowledge and commitment to achieve this. What will precipitate the shift is, however, totally unpredictable, and it is not my intention to explore this hypothetical scenario. Rather, I shall examine the shift in cultural values that are emerging as a result of the deep changes that have occurred in scientific culture during the twentieth century, largely in consequence of Chaos, Gaia, and Complexity Theories. These have resulted in an extended science that is no longer exclusively focussed on quantities and measurement as the ways to gain reliable knowledge of the world, with qualities and values entering again as significant indicators of balance and health in complex natural systems. Furthermore, there is a restoration of meaning as a major aspect of the living world and of evolution, replacing the existential emptiness of the mechanical world view. These are not just philosophical issues, but concepts at the heart of a new science.

Complexity and Qualities

The emergent properties of complex systems include qualities such as health and coherence of the whole, not simply quantifiable aspects of their dynamics. The human body is itself a complex whole with many remarkable properties including the condition of consciousness and the experience of feelings. The qualities of pain, joy, anger, depression and well-being are all qualities that are highly significant indicators of the condition of a person's body. Medical science has tried to avoid these as sources of information about the condition of a person's health, substituting measurable quantities such as temperature, blood pressure, blood cell counts, haemoglobin levels and many other physiological variables as indicators of health or disease. However, pain has never been ignored as a reliable sign by competent practitioners of any therapy, and the particular quality of pain experienced by a patient is very significant in evaluating the nature of the disturbance to the body. After being excluded from the set of observables that science allows as legitimate variables for the analysis of complex systems, which were restricted by the founders of Western science to measurable quantities, qualities are now returning as accepted evaluators of the condition of complex systems. A question that now arises is: what procedure or methodology is acceptable for demonstrating the reliability of qualitative evaluation for the condition of complex systems? The issue here is to define a method for a science of qualities. If we follow common sense, then we have no problem with this; everybody uses qualities as well as quantities in evaluating their experience of others and of their environment, cultural and natural. What some scientists tend to argue is that these are purely subjective evaluations that have no objective status, hence are not to be allowed as scientific evidence. However, there is now good evidence against this for many types of qualitative evaluation, and this is expanding.

I shall now describe one particular method for evaluating the qualities of complex wholes that is particularly relevant to the Gaia story. This procedure is known as Consensus Methodology, which uses a combination of Free Choice Profiling and a statistical method called Generalised Procrustes Analysis. The method was developed for the evaluation of the quality of food and drink, but has been extended to assessing the quality of experience of farm animals by human observers (Wemelsfelder et al., 2000; 2001). What this method demonstrates very clearly is that independent human observers reach a high degree of consensus about the quality of experience of farm animals such as pigs or cows without discussing their evaluations with each other. They simply look at the animals' behaviour and write down the words that convey what qualities of experience they think the animal is expressing, such as nervous, boisterous, excitable, laid-back, and so on. People select their words independently of each other and rate each animal on a quasi-quanti-tative scale of 0 to 1 for each word used. A computer programme does the rest, searching for clustering (consensus) in the collective space of all the independent observers and looking for the principal coordinates of the cluster. The results are remarkable in showing high degrees of consensus between observers whether they are farmers, vets, animal rights activists, or volunteers from any walk of life. We are good at evaluating the quality of life exhibited by animals through their behaviour. This is hardly surprising as we spend a lot of effort in evaluating the quality of experience of people with whom we are in relationship whether at home, at work, at play or anywhere else, by monitoring their behaviour in response to ours. Consensus Methodology is now being extended to the evaluation of the quality of other complex systems: of landscapes, rivers, ecosystems, and organisations. People have always used this as an important way of getting to know a place or a group, and people familiar with their environment are in a good position to evaluate any changes it undergoes, whether toward health or disease. Observers of the natural world have always believed they could tell when deleterious or beneficial change is happening through a combination of quantitative and qualitative examination of landscapes and communities. Their belief is now validated, and it is now possible for government monitoring services like the Environment Agency to use local stakeholders for this type of evaluation of their habitats. This is again good common sense, backed up by scientific method.

Such procedures for evaluating the health of both natural and cultural systems are just what is needed to get people involved in monitoring the quality of life that they experience in their organisations, in their communities and in their environments. What is beginning to happen here is an erosion of the sharp division that we have drawn between nature and culture. One of the primary objections to Gaia Theory was that Lovelock ascribed the properties of life to the Earth as a complex system because of its ability to regulate many of the variables crucial to the living state, as in physiological processes within organisms. This battle has been won, but there are still many who object to the cryptic animism of Lovelock's writing. What these objectors fail to recognise is that science has now developed to the point where complex systems with their emergent properties are seen to have intrinsic values of the type we recognise in conditions of health, coherence and well-being, and that these are now being used to describe observable conditions of complex processes, whether these are 'natural' or 'cultural'. In fact, I shall now argue that we have reached the point where words like 'meaning' can be legitimately applied to the processes in which natural systems engage, such as the life cycles of organisms or the maintenance of wholeness and healthy balance in ecosystems. We are reaching the point where, as Aldo Leopold put it, we can begin to 'think like a mountain'. I shall now articulate how biology has arrived at this point, so that evolution is no longer a meaningless mechanical process, but is full of meaning.

Meaning in Evolution

Biology conceived within the Darwinian perspective of life adapting to circumstance through random mutation in genes and selection of the fitter phenotypes through a process of trial and error was a triumph of mechanical thinking. It allowed for an unbroken causal chain from genes and the inheritance that are responsible for to progressively adapting phenotypes through the process of organismic self-construction (development) during life-cycles selected for adaptation and survival in different habitats. This accounted for both the diversity of species suited to different environmental niches and the progressive advance of evolution into ever more complex forms of life, a purely random exploration of possibilities. Twentieth century biology focussed on inheritance as the main vehicle of this process, identifying the double helix of DNA as the molecule that carried all the secrets of life. From 1953, the year in which Watson and Crick described how DNA structure allows for the twin processes of self-replication and of information coding for protein synthesis, until the end of the century biology was fully committed to unravelling the sequence of activities whereby these two key molecular actions of DNA achieved the self-reproduction of organisms, with inherited variations. This culminated in the successful readings of the sequence of bases in the genomes of different species, particularly the human, in 2001. However, this remarkable research achievement had a sting in its tail. Here is a passage from Evelyn Fox Keller's book *The Century of the Gene* (2000):

> What is most impressive to me is not so much the ways in which the genome project has fulfilled our expectations but the ways in which it has transformed them.
>
> Contrary to all expectations, instead of lending support to the familiar notions of genetic determinism that have acquired so powerful a grip on the popular imagination, these successes pose critical challenges to such notions. Today, the prominence of genes in both the general media and the scientific press suggests that in this new science of genomics, twentieth century genetics has achieved its apotheosis. Yet, the very successes that have so

stirred our imagination have also radically undermined their core driving concept, the concept of the gene. As the human genome project nears the realisation of its goals, biologists have begun to recognise that these goals represent 'not an end but the beginning of a new era in biology.

What is it that is emerging in place of genetic determinism? I now give an interpretation that comes from a perspective that questioned the primacy of genes even in 1953!

The issue that has always challenged the reductionist assertions of genetics, that genes are the necessary and sufficient conditions for life, is the question of the organism as a coherent whole. The processes described by molecular genetics and the genome projects are those involved in making the molecular parts of organisms, not how these are put together. It has been assumed that the myriad different molecules out of which organisms are made just get together and self-assemble into coherent wholes. It has been recognised that there needs to be a timed sequence of syntheses for this to occur during the development of an organism, but the genetic programme was assumed to contain all this information. However, one of the revelations of the genome projects is that the parts that biologists always believed to be the major constituents of organisms, namely proteins, account for less that 2% of the coding sequences in DNA. Most of the DNA was therefore described as 'junk'. It is this junk that has turned out to be crucially important in generating the coherent forms of organisms, by processes that are largely unknown. What is emerging is the realisation that the processes involved in producing coherence out of the molecular bits of organisms is achieved by networks of elements that interact with one another in very interesting and unexpected ways. This is the organised context that, somehow, makes coherent sense or meaning out of the information coded in DNA. Here we have the beginnings of a new biology that acknowledges processes of self-reference in the self-organising system of the cell. What is its nature?

Self-referential Networks and Language

Languages are self-referential networks of elements (words) whose meaning arises from their relationships with other words in the language. They have characteristic structures and properties. Can the networks of elements within cells, produced by genes, that are involved in turning the information in the genes into coherent patterns of agency and form, the living organism, be understood as languages? This is now a distinct possibility that is emerging from the detailed study of the ways in which these gene products act and interact. They have generic properties that are the same as those that characterise the frequencies of words in texts. Furthermore, these patterns can be shown to arise from a process in which the signallers and receivers within the network are achieving an optimum condition for creative communication that underlies a very important feature of language: the occurrence of ambiguity in the precise meanings of words. It is this ambiguity that is the creative resource of language, since it makes possible a diversity of forms of expression with subtly different meanings.

In organisms, the same type of ambiguity in the processes that are making the organism allows a great diversity of subtly different forms to emerge, each of which satisfies similar but not identical meanings, i.e., effective organisms in particular habitats. It is precisely the absence of mechanical causality in organisms, fixed patterns of gene action and determination, that allow for the continuous creativity of life. What is emerging in biology is a kind of equivalent of the indeterminacy revealed by quantum mechanics in the 1920s, when local causality had to give way to non-local connectedness and entanglement. These are holistic concepts in which part and whole are inextricably linked together through an ambiguity of state in the elementary particles that constitute the whole, defined by complementarity principles in quantum mechanics. The condition of coherence in quantum mechanics is described precisely by the same type of property that describes the frequency distribution of words in a language, defined as self-similarity or fractal structure. A metaphor for this is to say that the condition of coherence or wholeness gives maximum freedom to the parts *and* maximum order to the whole.

The equivalent of this in biology is the ambiguity of the elements in the self-referential networks that produce coherent organisms. They are not different from or outside the genes or the organism; they *are* the organism as a self-organising totality. And maximum freedom to the parts, maximum order to the whole, is achieved through the optimisation process that resolves the tension between signallers and receivers, the participants in the living conversation, by a principle known as least effort. The living state is the realisation of a condition of effortless effort, like the performance of an exquisitely fit and coherent dancer or athlete who achieves miracles of expression with minimal effort and maximum grace. Biology is now entering a period of flowering akin to what happened in physics about 100 years ago. It was believed that the paradoxes of the quantum world were restricted to microphysics, but now the living state is revealing its secrets and they are the same, though different. What is so important about this insight is that it shows why we will never be able to control organisms or manipulate their genes in predictable ways, as genetics has promised throughout the twentieth century. The very creativity of the evolutionary process depends upon the same kind of ambiguity of meaning that exists in language. Nature and culture now become one continuity of intelligible but not controllable creativity, combining freedom of the parts with coherent form of the whole. We seem to be recovering perennial wisdom in science.

Chaos, Gaia, and Eros

One of the expressions of perennial wisdom was known as the Orphic Trinity, the union of Chaos, Gaia, and Eros as the basic principles of endless creativity and freedom in the cosmos. Hesiod, the eighth-century Greek poet, tells us in his Theogony how Chaos, the yawning chasm of disorder, gave rise Gaia and Eros. These three powers or beings constitute the mysterious, magical forces underlying the whole of creation, Eros being expressed through the forms of nature. They were worshipped by the Pythagoreans, among others, as the Orphic Trinity that takes its name from Orpheus, the legendary musician, physician and spiritual leader who was also known variously as Dionysus, Osiris, Marduk, and Shakti in different traditions. And this Trinity of Father, Mother and Love was the basis of the Christian Trinity, though transformed by a patriarchal culture into Father, Son and Holy Spirit. Our culture now seems to be recovering a more balanced perspective on our relationship with one another and with the earth than that expressed in the Christian Trinity through a rediscovery of more appropriate ways of knowing and being than those developed in Industrial Culture.

I have mentioned how chaos and Gaia emerged from twentieth-century science during the 1960s. Another celebrated cultural transformation during that period was the emergence of the hippie life-style based on music and flower-power, an expression of Eros as Love in relationships. So the Orphic Trinity rose again from underground obscurity to transform the values of Western society. It has taken decades for this enlightenment to become grounded in practical movements towards life-styles that celebrate the same values as traditional cultures, but now based on the scientific and technological know-how we have developed to reduce our footprint on the planet and to become relatively invisible, like most other species. The task of achieving this has been described as *The Great Work: Our Way into the Future* by Thomas Berry (1999), and it is the challenge of our time. We shall either succeed in becoming integrated with natural principles or we shall not. As James Lovelock has graphically expressed it, Gaia as life on earth will survive whatever we do to the planet, with us or without us as a participant species. The transition to a sustainable life of quality and meaning from an unsustainable life of excessive quantities is a choice that is ours to make.

Sources

Berry, Thomas (1999), *The Great Work: Our Way into the Future* (New York: Bell Tower).

Keller, Evelyn Fox (2000), *The Century of the Gene* (Cambridge, MA: Harvard University Press)

Stephan Harding

Animate Earth

There is now virtually no doubt that our culture has unleashed a massive crisis upon the world. We are changing the very climate of the Earth, we are driving millions of species into extinction and we are eroding the social ties that bind us into healthy families and communities. There are many metrics that can help focus our minds on the immensity of what is happening—here are just two. Firstly, because of our burning of fossil fuels, the Earth has not experienced atmospheric carbon dioxide levels as high as today's for about 740,000 years, and secondly, we are wiping out our fellow species with a ruthlessness that beggars the imagination—every day we exterminate some 100 species around the world at a rate about 1000 times faster than the natural background rate of extinction. The crisis is so colossal that some eminent scientists, such as James Lovelock, warn that we could face the collapse of civilization within a matter of decades.

It is now widely accepted that we have brought this dangerous predicament upon ourselves through our pursuit of unlimited economic growth powered by the burning of fossil fuels and the ruthless exploitation of wild ecosystems. But there is a deeper cause that lies not in these outer actions, but in the very way that we have been taught to see the world ever since we were children. This is a worldview so dangerous that it has led us to wage an unwitting war on nature that we cannot possibly win, a view that we must quickly modify if we are to have any viable future on the Earth.

Our death-dealing worldview is simply this: that, for us, our great turning world is no more than a vast dead machine full of 'resources' that have value only when they are converted into money. We think that mountains, forests, and the great wild oceans are all dead things that we are free to exploit as we wish without let or hindrance. Our culture values only quantities such as weight, height, and money in the bank. We have been taught to disregard qualities—to believe that the sense of elation we feel in the mountains or the calm we experience by a sunlit lake are merely our own idiosyncratic subjective impressions that tell us nothing real about the world. Furthermore, we have been persuaded to think that good citizenship involves buying more and more material goods so that the global economy can grow.

But what if our relationship to nature is dysfunctional because of an unbalanced psychological development both individually and within the culture as a whole? And if this is indeed the case, is there anything we can do about it? It was C.G. Jung, the great Swiss psychologist, who observed that we all have four psychological functions, or 'ways of knowing', which operate as pairs of opposites: Intuition and Sensing, Thinking and Feeling. Intuition gives insight into the nature and deeper meaning of things, whilst sensing yields a direct apprehension of the world around us through the substrate of our physical bodies. Thinking interprets what is there through the exercise of logic and reasoning, whilst feeling helps us to ascribe positive or negative value to phenomena and situations — ultimately this is the sphere of ethics. Thinking and feeling are evaluative, whilst sensation and intuition are perceptive. Jung discovered that each of us has a dominant function, whilst the opposite function remains largely unconscious and undeveloped. The other two functions are only partially conscious, generally serving the dominant function as auxiliaries. Of course, this typology suffers from the limitations of all models, but Jung found it useful enough to say of it that it 'produces compass points in the wilderness of human personality'. Mental and physical health in Jung's therapeutic approach requires the conscious development of the neglected function together with an awareness of the four functions in oneself so as to achieve a well-rounded personality.

By applying Jung's insight to our culture, it is easy to spot the fact that we are suffering from an over-development of a particularly dangerous style of thinking that became immensely persuasive during the scientific revolution in the sixteenth and seventeenth centuries. It was at this time that philosopher/mathematician René Descartes and others made a convincing case that there is a fundamental ontological divide between the rational human soul and a soulless material universe, which was seen as nothing more than a vast mechanical contrivance there for us to dominate and control with impunity through the exercise of pure analytical reasoning.

This mechanistic worldview, useful as it is in certain limited ways, is at odds with a more ancient sensibility that saw the Earth and the entire cosmos as sentient beings worthy of reverence and respect. Our ancestors, and indeed many indigenous people to this day, sensed that they lived within a great psyche, the psyche of the cosmos itself — the *psyche kosmou*, as Plato called it, or, in Latin, the *anima mundi* — the 'soul of the world'. In this view, matter is sentient to its deepest roots — everything is capable of experience, including subatomic particles, mountains, entire ecosystems, and indeed the Earth itself. Father Thomas Berry brilliantly expresses this insight when he says that 'the world is not a collection of objects but a communion of subjects'. According to this 'panpsychist' or 'animistic' perspective, our own consciousness has its roots within the atomic mode of sentience, for it must somehow emerge out of the complex and intricate communion that takes place amongst the atoms that make up our physical bodies. But we can only become truly sensitive to the living qualities of the world when we combine these rational arguments with our intuition, sensing and feeling. These tell us that every speck of matter has intrinsic value irrespective of any use that we might put it to, and that nature's subtle qualities — her colours, sounds and textures — are a kind of language or 'text' that constantly speaks to us of our animate surroundings.

Strangely enough, the idea that the Earth is alive has come back to the modern world in an unlikely arena — within science itself. In 1972 James Lovelock proposed that our planet consists of a tightly coupled set of complex feedbacks between life, rocks, air and water that gives rise to the emergent ability of the planet as a whole to regulate its own surface conditions over past periods of time within the narrow limits suitable for life. Inspired by William Golding, Lovelock chose to name his theory of a self-regulating Earth after Gaia, the ancient Greeks' animistic divinity of the Earth. A key insight from Gaia theory for us to ponder is that we humans are not in charge of the planet — that we are not the most important of her species. Instead, in the words of Aldo Leopold, the influential American environmentalist of the last century, we are 'just plain members of the biotic community' — special in our own unique ways, but no more special in principle than the trees, the great whales, the teeming denizens of the microbial realm or the millions of other species that populate our newly stricken world.

Perhaps it is time to counter our dangerously outdated mechanistic worldview with a more fruitful, more soulful, science-based idea in tune with the wisdom of our ancestors that inspires us to uncover our deep indigenous connection with the earthly community of rocks, atmosphere, water and living beings — with the animate Earth that enfolds us. We can do this by learning how to relate holistically to the animate Earth through our four ways of knowing. We use our reason to study the Earth as an emergent self-regulating system that has kept our world habitable since the appearance of life some three and half thousand million years ago thanks to the tumultuous and multitudinous interactions between living beings and the atmosphere, rocks and water that surround them. This is the essence of Gaia theory, which teaches us that we live symbiotically within a vast evolving sentient creature that has been charting her yearly path around the sun for thousands of millions of years, evolving her capacity for keeping her crumpled surface suitable for life as her biodiversity has increased over geological time. We can ponder the consistencies or otherwise between Gaia theory and the theory of natural selection, we can build mathematical models of the carbon cycle coupled to an active biota, and we can look at how the Earth could respond to climate change as a fully integrated complex system consisting of life coupled to its abiotic environment. But we speak of the players in this 'system' not as dead cogs in a static machine, but as animate beings with particular and oftentimes peculiar personalities. Carbon atoms are the placid Swedes of the chemical world; oxygen atoms are its passionate Italians — for if the world is truly ensouled then even atoms are 'persons' in the most rudimentary sense of the word.

Then we go further. We can use this rational knowledge to fuel our intuitive sense of connection to the whole community of nature by recreating Gaia's long and complex evolutionary trajectory in our imaginations and by engaging in rigorous meditative explorations of her tightly coupled feedbacks. We deliberately connect with the qualities of rocks, atmosphere, oceans, clouds, individual organisms and entire ecosystems by spending quiet time savouring their essences much as we would that of a poem or a piece of music. As we deepen our perceptual abilities, we find a remarkable degree of agreement with each other in what we discover by means of this more phenomenological approach to nature. In addition, we work with exercises that help to shift our everyday perceptual frameworks.

I offer you one such 'meditation'. Just try it. I guarantee that it will give you an unexpected depth of belonging to the Earth. From the point of view of our mainstream culture you will be 'doing nothing', but in fact you will be engaging in highly subversive act — the demolition within yourself of our suicidal and vastly destructive mechanistic worldview.

> *Lie on your back on the ground outside in as peaceful a place as you can find, in the forest perhaps, or by the roaring sea. Relax and take a few deep breaths. Now feel the weight of your body on the Earth as the force of gravity holds you down.*
>
> *Experience gravity as the love that the Earth feels for the very matter that makes up your body, a love that holds you safe and prevents you from floating off into outer space.*
>
> *Open your eyes and look out into the vast depths of the universe whilst you sense the great bulk of our mother planet at your back. Feel her clasping you to her huge body as she dangles you upside down over the vast cosmos that stretches out below you.*
>
> *What does it feel like to be held upside down in this way – to feel the depths of space beyond you and the firm grip, almost glue-like grip of the Earth behind you?*
>
> *Now sense how the Earth curves away beneath your back in all directions. Feel her great continents, her mountain ranges, her oceans her domains of ice and snow at the poles and her great cloaks of vegetation stretching out from where you are in the great round immensity of her unbelievably diverse body.*
>
> *Sense her whirling air and her tumbling clouds spinning around her dappled surface.*
>
> *Breathe in the living immensity of our animate Earth.*

Do this again and again, at every available opportunity. Let yourself be '*Gaia'ed*' by the great round sentience of our living world. Deeply experience what it feels like to meld with the great wild body of our animate Earth in this way. See how this simple act brings you into a deeply felt relationship with the whole of life.

Experiences such as these lead us into the realm of ethics as we deeply question our own lifestyles and those of our society in the light of our deepening connection with the personhood of the Earth – this is the 'deep ecology' approach of the great Norwegian philosopher Arne Naess. When we do this, we sense that it is wrong to seriously harm the great turning world within which we live. This realisation gives us the energy and insight to change our lifestyles in beneficial ways. Research by people such as David Reay at the University of Edinburgh has shown that we can make a massive difference in our personal lives thanks to simple acts such as: turning down our heating in winter by just 1 degree centigrade; using our cars less; composting organic waste; avoiding flying; driving at or below speed limits; eating locally produced food; reducing, reusing and recycling; and by turning off all standbys and transformers. Also, we can think carefully before we buy anything new. Could we buy it second-hand, or even do without? We can become involved in strengthening our local communities, and find satisfaction in talking, telling stories and making music together rather than in working so mindlessly hard to buy the mostly useless consumer products promoted by the mass media for filling the gaps in our lonely lives. All of this doesn't seem like much, but if enough of us consume less in these ways we will make a huge difference, thereby removing the need for several new power stations in the UK.

Connecting with the Earth, consuming less and developing a sense of community can also give us the will and energy to work for change at the societal level. In this domain perhaps the most important thing to do is to agitate for an economy that is in a steady state, rather than working blindly for one which seeks to grow by extracting more and more of the Earth's finite resources from her ancient crumpled surface. Those of us touched by the animate Earth feel the urge to work towards creating an economy in which the things that grow are the development and deployment of renewable technologies, the restoration of degraded ecosystems, the recreation of vibrant local communities and economies, and the adoption of ecologically diversified farming practices. Policies inspired by this kind of 'intelligent growth' would also stimulate those non-material things that can grow without limit — spirituality, creativity, depth of community and simple living. These are, after all, the sources of our deepest satisfactions and of our sense of well-being.

So how can we promote intelligent growth? We can begin by consuming less in our personal lives, as outlined above. But we can also lobby government (via our MPs) to take climate change seriously by setting a ceiling on greenhouse gas emissions through the implementation of a rigorous carbon rationing system, by giving significant tax breaks and other incentives for implementing energy-saving measures, by funding massive research efforts into renewable energy and by developing ecologically sound ways of food production, building and transportation.

If we don't immediately make these radical changes we will almost certainly invoke the fearsome wrath of Gaia. Her law is that any being that destabilises her climate will experience feedbacks from the whole 'system' that will curtail the activities of that being. So we have a choice. We can carry on with business as usual and live in rightful fear of Gaia. Or we can learn to love her hills, her wild forests and her oceans as we love a cherished grandmother. Perhaps only then, motivated by this love of all earthly things, will we find the inspiration for mending our ways and for massively reducing our impact on the great animate being that gave us birth.

Sources

Abram, D. (1997), *The Spell of the Sensuous* (Vintage).
Berry, T. (2000), *The Great Work* (Crown Publications).
Harding S.P. (2006), *Animate Earth: Science, Intuition and Gaia* (Green Books).
Jung. C.G. (200), *On the Nature of the Psyche* (Routledge).
Lovelock, J.E. (2006), *The Revenge of Gaia* (Penguin Books).
Lovelock, J.E. (2005), *Gaia: Medicine for an Ailing Planet* (Gaia Books).
Sessions G. (ed), *Deep Ecology for the 21st Century* (Shambala).
Skrbina, D. (2005), *Panpsychism in the West* (MIT Press).
Wilson, E.O. (2002), *The Future of Life* (Little, Brown).

Cormac Cullinan

Gaia's Law

James Lovelock has described how he first glimpsed Gaia in a flash
of enlightenment in Pasadena, California in the autumn of 1965. He
had been wondering why the composition of the Earth's atmosphere
had been constant over long periods despite the fact that the mix of
gases was chemically unstable. The insight came when he asked
himself if the explanation might be that life on Earth was somehow
regulating the atmosphere and keeping it at a level favourable for
organisms.[1] The question of self-regulation has been central to the
Gaia story ever since.

Probably the key question that the current generation must
answer is, 'What role will we choose to play in Gaia's self-regulatory
system?' The now readily observable phenomenon of climate
change imbues this question with a particular urgency and gravity.

For about 3.8 million years the Gaian system has regulated itself in
such a way that the composition of the atmosphere and the average
surface temperature of the planet have been maintained at levels
conducive to the evolution of life. It has done this despite the fact
that the luminosity of the Sun has increased by 25% during this
period.[2] However since the industrial revolution humans have been
digging up and burning large quantities of fossil fuels that contain
carbon removed from the atmosphere by living organisms. In addition to accelerating emissions of carbon dioxide and other greenhouse gases, we have simultaneously been destroying forests and
other ecosystems. This has impaired Gaia's ability to restore the balance by increasing the rate at which these gases are sequestrated. It is
now clear that humans have impaired the functioning of Gaia's
self-regulatory systems to such an extent that they are no longer
capable of keeping the average temperature of the planet within the
range that has been maintained for millions of years, (or at least not

---

[1] Lovelock (2000), pp. 21–22.
[2] Ibid p. 22.

within timescales that are meaningful to humans). The extent to which we continue doing so will be determined largely by the functioning of our governance systems (i.e. policies, laws, institutions and the philosophies and values that inform them). This means that the functioning of human governance systems will be an increasingly important determinant of the functioning of Gaia's self-regulatory system. Accordingly it is essential to apply a Gaian perspective in understanding and urgently reforming the laws and other self-regulatory mechanisms of our societies.

Getting to grips with the implications of the Gaia theory for law and governance requires an appreciation of the extent to which we humans are part of Gaia and consequently both influence, and are influenced by, the functioning of the system as whole. We often forget that we are part of an immense creative story that has brought us, and all who inhabit Gaia, to this momentous transition in her evolution. Since the Big Bang, the Universe has displayed an awesome and apparently endless capacity to differentiate itself into an astounding array of self-organising forms, which continue to maintain relationships with one another in a way that maintains the integrity of the whole. The unfolding story has seen the creation of the elements, galaxies, solar system, and Gaia herself.

This opalescent, milky-blue planet has proved to be uniquely creative. She has brought forth a myriad of carbon-based life forms that have bound themselves through a web of relationships into intricate natural communities of stunning diversity and beauty. For most of our existence we humans have had little impact on the vast Gaian systems that regulate the composition of the atmosphere and the climate. However it now appears that our particular gifts of self-reflective consciousness, intellect and imagination, have the potential to play a decisive role in Gaia's evolution in the foreseeable future.

Our consciousness enables us to understand that we are part of an on-going evolutionary story, gives us insight into the functioning of the system as a whole, and a degree of choice as to what happens next. The golden aspect of these prodigious gifts is revealed in our sense of beauty and the wonderful art, music, languages and cultures that humans have produced. Its dark aspect is manifest in the use of our intellect to set ourselves apart from Gaia and to seek to dominate and exploit nature. The arrogance and chauvinism of contemporary humans in relation to Nature often blinds us to the fact that we owe our existence and continued survival to the community within which we evolved. Everything about our species, from the size of our brain, the form of our bodies, to our sense of beauty and to colour has been shaped by an intimate dance with the plants, animals and microbes with which we have co-evolved.

Our contemporary alienation from Gaia and our loss of reverence for the other species and natural processes that sustain us all wouldn't matter much if we didn't have such a great capacity to alter the planet. For aeons natural selection was the dominant force shaping life. Now the fate of most species is being determined by conscious human choices. How we use our imaginations, develop our cultural understandings and apply our technological expertise will exert a perceptible influence on Gaia's unfolding life story. There is no time left for natural selection to guide the majestic shaggy bears of the Arctic to a new ecological niche. Unless humans make radically different choices very soon, the long saga of the great yellow-white bears and their Arctic habitat will end abruptly, impoverishing all future generations.

An alien visiting Gaia at this juncture in her history would probably not wager much on humans changing their behaviour fast or radically enough to avert massive damage to ecosystem and the death over an extended period of most, possibly all, of the human species. However, as intelligent Earthlings who must bear a great deal of responsibility for the destruction of other life forms and habitats that is currently occurring, it is incumbent upon us to focus our immense gifts on re-aligning our governance systems. We need to change the laws that define all non-human aspects of Gaia as property, which by definition cannot have rights. Laws that encourage the establishment of legal-fictions such as companies in order to facilitate the ruthless exploitation of Gaia, while protecting those responsible from financial and psychological responsibility for the social and ecological harm that they cause, are part of the problem and must be reformed if we are to repair the harm we have caused.

If we are to re-orientate human behaviour so that it benefits Gaia, we must first abandon our cherished delusions that we are separate from Earth, superior to other beings and capable of indefinitely enhancing our well being at the expense of Gaia. Despite our illusions, the truth is that, from a systems perspective, we are merely an aspect of the 'super-organism' or system that we call Gaia. Physically our bodies are derived entirely from hers, and her beauty, smell, taste and texture have shaped our senses, formed our minds and inspired our creativity. We cannot sustain our existence independent of her. What befalls Gaia will also befall us, and the harm that we inflict on Gaia we ultimately inflict on our own kind.

Once we again see ourselves as members of the whole Earth community and re-awaken our cultures to the Gaia system within which they are embedded, the need for legal and political philosophies that reflect this Gaian perspective becomes obvious. Current legal philosophies (jurisprudence) based on notional social contracts between humans does not provide an adequate basis for developing legal and political systems that reflect the reality that our societies are an integral part of the Gaian system. In order for us humans to develop mutually enhancing, rather than exploitative, relationships with Gaia we need an Earth jurisprudence that explains why the fundamental purpose of our governance systems must be to regulate humans so that they contribute to, rather than undermine, the health and flourishing of Gaia in all her manifestations.

The term 'Earth jurisprudence' refers to the philosophy of law and human governance that is based on the belief that human societies should regulate themselves as members of a wider Earth community and in a way that is consistent with the fundamental 'laws' or principles that govern how the universe functions and are not determined by humans.[3]

For example, the cosmologist Brian Swimme and the cultural historian Thomas Berry refer to 'the Cosmogenetic Principle' that postulates that the evolution of all parts and dimensions of the universe will be characterised by three qualities or themes: *differentiation, autopoiesis* (meaning literally 'self-making') and *communion*.[4] 'Differentiation' refers to an inherent tendency towards diversity, variation and complexity; 'autopoiesis' to an inherent ability to self-organise and to be self-aware, and 'communion' to the interconnectivity of all aspects of the universe. In short, the universe orders itself by differentiating itself into recognisable different aspects or parts, each of which orders and regulate itself internally (autopoiesis),[5] and is organised in relation to all other aspects (communion).

---

[3] The author of this article discusses these concepts in more detail in *Wild Law* (2003). See also Berry (2006), particularly chapter 9 'Legal Conditions for Earth Survival' and Appendix 2 'Ten Principles for Jurisprudence Revision'.

[4] Swimme & Berry (1994), pp 73–75.

[5] Each self-organised part is a whole. As Jan Smuts observed in 1926: 'Both matter and life consists of unit structures whose ordered grouping produces natural wholes which we call bodies or organisms. This character of "wholeness" meets us everywhere and points to something fundamental

Furthermore, Gaia is an evolving system from which new characteristics or properties emerge as it increases in complexity. As long ago as 1926, Jan Smuts (1870–950) postulated that inorganic matter produced life, which in turned produced mind, and that each superior level is greater than the sum of the lower levels and could not be reduced to its constituent parts. Furthermore, natural systems exhibit what Edward Goldsmith calls 'whole maintaining' characteristics.[6] In other words, each aspect of a properly functioning system acts in a manner that contributes to the health and integrity of the whole. If this were not the case then the whole system would begin to deteriorate with adverse consequences for all the component parts. This means that any organism or community that is unable to regulate itself in a manner that ensures that its component parts or members function in a way that benefits the whole, will ultimately disintegrate.

A consideration of these fundamental characteristics of the universe immediately suggests that if human governance systems are to be consistent with Gaia's regulatory system, they must be concerned with maintaining and strengthening relations between all members of the Earth Community and not just between human beings. In other words they must serve a 'whole maintaining' function. It also suggests that we should reconsider our preoccupation with achieving uniformity at the expense of cultural, biological and other forms of diversity.

If we respect the principle of diversity, it follows that each different human community may have a different view on how best to regulate itself as part of Gaia. Accordingly there are likely to be a variety of Earth jurisprudences. Indeed this is borne out by the many different cosmologies of indigenous communities that seek to live in accordance with what they perceive to be the timeless laws that govern their world (and therefore exhibit aspects of Earth jurisprudence). However, since each variation of Earth jurisprudence must be shaped to accord with the natural laws that govern the system as a whole,[7] all variations are likely to share common elements. For example, I anticipate that any Earth jurisprudence would:

- recognise that it exists within the wider context of Gaia that shapes it and influences how it functions;

---

in our universe. Holism ... is the term here coined for this fundamental factor.' See Smuts (1926).

[6]   Goldsmith (1992).

[7]   I have referred to this elsewhere as 'the Great Jurisprudence' (2000).

- recognise that non-human members of Gaia also have fundamental 'rights', such as the right to existence and habitat and
  the freedom to play their role in the great evolutionary story,[8]
  and that humans have a corresponding duty to ensure that
  human actions do not unjustifiably infringe on those rights;

- regard the universe, rather than human legal systems, as the
  source of these fundamental rights and accordingly regard any
  attempt by human jurisprudence and laws to seek to circum-
  scribe or abrogate them, as illegitimate;

- reflect a concern for reciprocity and the maintenance of a
  dynamic equilibrium between all the members of the Earth
  Community determined by what is best for Gaia as a whole
  (Earth justice); and

- condone or disapprove of human conduct on the basis of
  whether or not the conduct strengthens or weakens the bonds
  that constitute the Earth Community that is Gaia.

It is also important to appreciate that this approach has both a personal or internal dimension and an external social dimension. Since
we are part of Gaia both as individuals and as societies, applying
Earth jurisprudence requires both that we each internalise these
insights and change our personal ethics and practices, and that we
develop policies, laws, and institutions that embody it.

Gaia theory is particularly useful in helping us to understand that
we are an integral part of a greater system and that in order for our
legal and political systems to reflect this reality it is necessary to consciously shift our reference point away from a purely human-centred one. The shift of perspective that is required is analogous to the
shift of mental perspective that was required in order to appreciate
the validity of Copernicus's discovery that Earth revolved around
the sun and not *vice versa*.

It is also useful to remember that one should expect resistance to
such a fundamental shift. In the case of Copernicus and Galileo the
authorities of the time so feared the consequences of accepting that
Earth, and ultimately human beings, were not the centre of the universe around which everything else revolved, that they attempted to
discredit the theories and forced Galileo to recant in public. In doing
so they appear not to have realised that this would have no effect on

---

[8] Thomas Berry, a leading cultural historian, 'geologian', philosopher and
writer has formulated the basis of what he considers these rights to be. See
for example, 'The Origin, Differentiation and Role of Rights' published in
Cullinen (2003) at page 115, and Berry (2006), pp. 149–150.

reality, and would merely perpetuate an illusion. Similarly, although the idea of governing ourselves as if we were an integral part of Gaia may seem radical, and even threatening, to some, it amounts simply to recognising a reality. The choice is not between adopting this approach or some equally effective alternative governance system. It is between either advocating the application of Earth jurisprudence or the perpetuation of governance systems based on premises that we now know to be false, and that are demonstrably harmful. The choice is between Earth realism and denial of the worst and fatal kind.

However we must also not fool ourselves into believing that a mental shift of perspective and a change of heart is all that is required. Preventing further destruction of the Gaian community and repairing some of the harm done will require us all to roll up our sleeves and devise and implement new governance systems. Systems that will inspire, cajole and compel humans to respect the natural laws and limits to which we are all subject.

We need to devise 'wild laws' that recognise and embody the qualities of the Gaian system and that seek to foster intimate and healthy relationships between humans, and other aspects of Gaia such as animals, plants, rivers and forests. Laws that re-define terms like 'anti-social behaviour', 'social responsibilities' and 'justice' in relation to the whole Earth community and not just in relation to humans. Laws that require humans to honour their responsibilities to those other beings in whose company they co-evolved.

This is unlikely to be an easy task. Today most humans are so autistic to the presence of Gaia that we are no longer conscious that they are part of Gaia and subject to the principles or 'laws' that characterise her self-regulatory system. However the very process of developing governance systems according to 'whole maintaining' criteria will begin the process of eliminating self-destructive anomalies in the system, help to create more integrated communities, and ultimately help restore integrity to Gaia's self-regulating systems. Re-orienting our societies in this way certainly presents an immense psychic, intellectual and physical challenge — but one that just might be exciting, challenging and dangerous enough to inspire the flowering of a new kind of civilisation.[9]

---

[9]   Fortunately we can already see small shoots emerging in various forms, including the establishment in 2006 of the first Earth Jurisprudence Centre in Florida as a collaborative venture by the Universities of Barry and St John, the holding of regular conference on Wild Law in the United

Sources

Berry, Thomas (2006), *Evening Thoughts: Reflecting on Earth as Sacred Community*, ed. Mary Evelyn Tucker (San Francisco: Sierra Club Books).

Cullinan, Cormac (2003), *Wild Law: A manifesto for Earth Justice* (Dartington: Green Books).

Goldsmith, E. (1992), *The Way: An Ecological World View* (London: Rider Books; revised edition printed by Green Books in 1996).

Lovelock, James (2000), *Gaia: The Practical Science of Planetary Medicine* (Gaia Books Limited).

Smuts, J. (1926), *Holism and Evolution* (quoted in D. Pepper, *Modern Environmentalism. An Introduction*, London, Routledge, 1996).

Swimme, Brian & Berry, Thomas (1974), *The Universe Story* (HarperCollins).

Kingdom, and the dissemination of similar ideas through the African Biodiversity Network, and the Living Democracy movement in India.

John Turnbull

Can We Get There From Here?

Why the Gaian Worldview Will Struggle

The Gaian worldview will not spread far under current conditions. A worldview that recognises the interconnectedness of all aspects of life, and hence the need to intervene in systems to keep them in balance, is up against a political-economic system in which powerful vested interests, marching under a banner of spurious economic ideology, act in ways that show little care for that balance.

We live in a world system that depends not upon balance but upon division: division of actions from consequences and of power from real democratic scrutiny. These divisions are not the natural consequences of human nature, but are artificial: emergent properties — and, at the same time, necessary conditions — of a particular social/economic/political system.

Where the Gaian worldview sees society as a sub-system of the larger Gaian system, the dominant economic ideology sees nature as something outside society and the economy — a pool of resources to be exploited, with no fear of depletion. Where the Gaian worldview recognises that human actions have far-reaching consequences that must be accounted for, the increasingly globalised economy places an ever-greater distance between one person's decision to act and the consequences of that action.

Whatever its flaws, however, this system has served a minority of the planet's population extremely well. The Western world enjoys a material standard of living that was unimaginable even just a few decades ago. But the majority has been left behind. Immanuel Wallerstein argues that this system has left more people worse off —

in *absolute*, not just relative, terms — than ever before (Wallerstein, 1996).

Why does this system persist, when it serves most of those living within it so badly? Even the 'winners' — those of us in the rich world who benefit from it — are beginning to accept that there are major, potentially catastrophic, flaws in it: rarely a day goes by, for example, without news of global warming or the increasing number of migrants going to desperate lengths to escape the poor world for the rich.

We are beginning to join the dots and recognise the interconnectedness, but still very little is done to intervene in the system at the appropriate points. If our neighbour up the hill had a leaking pipe that was flooding our front yard, we know that we would not try to solve the problem by building a waterproofed wall around the yard; instead we would ask the neighbour to fix the pipe.

With larger scale problems, such 'common sense' seems to desert us. We instigate ever more draconian immigration laws, for example, rather than working to make the global system fairer and hence migrants' home countries better places to live.

The question of why this system persists seems simple, but the issues are complex and messy. There is no widespread agreement on what the problem is, much less the solution. However, there are two important and interlinked factors that help to sustain the current system and hence prevent the widespread take-up of a Gaian worldview: vested interest and economic ideology.

Vested Interest

It's easy enough to point to examples of vested interest amongst the ranks of the powerful: oil companies, for example, either denying the existence of global warming, or accepting that it is happening but denying that burning fossil fuels contributes to it.

'Big business' as a whole has good reason to resist major change to the configuration of the current system, especially after the last few decades of economic globalisation and soaring executive pay packages. The leaders of global corporations are not evil individuals, but are working within a system that demands short-term gain and rewards them very well for achieving that. (It also happens to reward them very well for *not* achieving.)

Vested interest in the current system is of course not limited to corporations. If it were, things would be simple: governments could intervene at the appropriate place in the system to improve matters (if not solve problems altogether). Were a connection to be made, for example, between economic globalisation and environmental depredation, governments might amend company law to change the way corporations operate, perhaps by requiring them to take full responsibility for all their external costs, including damage to the environment.

That this doesn't happen—or happens, at best, in a half-hearted way—is in large part due to the vested interests of political systems. Such action would drive corporations abroad to countries with laxer business environments, 'upset' the financial markets and play very badly in the corporate media; votes would almost certainly be lost. There is little prospect of change in the short- to medium-term: with many governments looking to private companies to deliver public services, the line between the interests of politics and business is increasingly blurred.

If the governments have a vested interest in keeping big business happy, so too do political parties and individual politicians. Parties in the UK, for example, increasingly rely on donations from wealthy benefactors, as party membership dwindles; and there is now, more than ever, a revolving door between politics and business, as former ministers and senior civil servants are hired by companies eager to make use of their contacts when bidding for government contracts.

So much for business and politics: nobody in these cynical times would be too surprised by anything we've seen so far. But what about the rest of us? If we see big business and economic globalisation as part of the problem, if we see the hidden connections, why don't we vote with our feet?

The fact is that almost all of us in the rich world contribute to the profits of trans-national corporations and therefore to their continued existence; we like our lifestyles and we won't easily be persuaded to change. Even if we decided to reduce our dependence on 'big business', short of downsizing to a hunter-gatherer existence, it's impossible to avoid altogether.

The vested interest of most people (i.e. all but the poorest) in the rich world is a huge barrier to the spread of a Gaian worldview. We see the TV news, we read the newspapers: we make connections and we know, deep down, that our lifestyles are not sustainable, but we don't really want to change; life is very comfortable, thank you. The long-term effects of that lifestyle—climate change, for example—are difficult to comprehend from within the framework of a system focused on short-term gain: quarterly profits and elections every four or five years. Living in a short-term culture, it's not surprising we can only think in the short term.

So vested interest, whether of powerful corporations, governments or people in general, is an important factor preventing us from taking action to intervene in our economic and political systems to help resolve some of humanity's most pressing problems. It is, in other words, a barrier to Gaian thinking.

But vested interest is rarely naked: there is almost always some cloak of respectability, a means of justifying our actions. In the case of people in general, it is usually that we are not doing anything that our neighbours aren't: if everyone else flies three times a year, why shouldn't I?

The most important 'cloak', however, is that which covers our entire economic system: neoliberal economic ideology.

Economics As Ideology

The second important factor preventing the spread of a Gaian worldview is the power of the economic ideology used to justify the workings of the current economic system. This ideology is largely propounded by those with the most to lose from any change in the system: those with the greatest vested interest.

The true nature of our economic system and its inter-relationship with our political systems is obscured by false claims and wishful thinking. Seeing our economic system for what it is is further hampered by the fact that we all live in it every day of our lives, rendering it almost invisible; we can't see the wood for the proverbial trees.

The problem lies not with economics as an academic discipline but with the way that economic theory is used and abused. Economics as a field of study is concerned with 'the production, distribution, and consumption of goods and services and [...] the theory and management of economies or economic systems' (*American Heritage Dictionary*). Economists, like other social scientists, study real-world social phenomena and try to provide an account of it; or at least that's the theory.

In reality, economists in influential positions outside academia — in think tanks, businesses and in international organisations such as the World Bank and the IMF — tend to champion a particular subset of economic theory; not because it accurately reflects the way the economy works, but because it fits with their ideological predispositions and can be employed to justify the workings of our current system. Governments, not surprisingly, given their vested interest in keeping the business world happy, buy into this.

Moreover, this ideology is promoted as being not an ideology at all, but a given truth. According to John McMurtry (quoted in Rees 2002):

> [Like] other social value programs, the doctrine of 'the global free market' itself does not recognize its ideology as ideology, but rather conceives of its prescriptions as *'post-ideological'* recognition of law-like truth [original emphasis] … The truth of the global market order is believed to be final and eternal, 'the end of history'. Its rule is declared 'inevitable'. Its axioms are conceived as 'iron laws'. Societies that dare to evade its stern requirements are threatened with 'harsh punishments' and 'shock treatments' (McMurtry, 1998, p. 43).

If this system seems not to work for some of us, tough: instead of adapting it to better meet the needs of all, the prescription is for *us* to adapt by becoming the type of economic agent this theory requires: *Homo economicus*. However, as William Rees, says (ibid):

> Like all abstractions, the global market model/myth simplifies reality — for example, it transforms decent well-rounded citizens into gluttonous single-minded consuming machines. The resultant *Homo economicus* is defined as a self-interested utility maximizer with immutable preferences and insatiable material demands (definitely not the type of person one might invite home to dinner!). You and I are assumed to act as isolated automatons whose sole goal is to maximize our personal consumption through participation in the increasingly global marketplace. The market model cannot accommodate the concept of 'family' and relieves our morally diminished *Homo economicus* of any other responsibility to society … .

This economic ideology is often called 'neoliberalism' and can be described, in a nutshell, as the belief that there should be as few restrictions as possible on the operations of businesses and the economy in general; that government should privatise all but its core functions, such as defence and the judiciary; and, at the international level, that overseas markets should be opened up by the use of political, economic or diplomatic pressure.

Underpinning all this is the view that 'so long as market pricing is in place and technological innovation continues, economic growth cannot be derailed by scarcity. For all practical purposes, natural resources are infinite' (Gray 2004, p. 59) (It hardly needs pointing out that this does not accord with the Gaian worldview.)

Free markets are held to be the most efficient way for nations to run their economies, especially since the collapse of communism (which very few people would argue was a preferable system). At the national level, areas once outside the marketplace (at least in western Europe), such as health and education, are increasingly being opened up to 'market forces'; at the international level, global free trade is promoted as the cure for the developing world's ills.

Taken at face value, the arguments for each of these courses seem persuasive. Private companies, with their responsiveness to market forces, are much more efficient than central planning by governments, so why not let them inject some of that efficiency into, for example, the National Health Service? And who are we to deny the developing world the opportunity to become as rich as we are, by adopting the free trade model?

But it's not so simple. First, while some areas of the economy, such as consumer goods, lend themselves naturally to a market system, others, such as healthcare, do not. Consequently they have to be artificially manipulated and moulded so that market principles can be applied, usually with poor results. Institutions, like people, must be made to conform to the neoliberal view of how the world works. No matter what politicians say (and some perhaps even believe) about providing greater 'choice', the marketisation of areas once considered off limits is primarily for the benefit of corporations.

Second, what is advertised as being the free market very often is not. The neoliberal belief is that absence of regulation equates to economic freedom; in reality, a truly 'free' market, in which many firms compete on a level playing field, requires robust government regulation. An absence of appropriate regulation results not in freedom but in the law of the jungle, with the most powerful corporations not competing but working actively to eliminate all competition in order to monopolise markets. Genuinely free trade, properly regulated, might well benefit developing nations, but the terms set by the neoliberal prescription are simply that these countries be opened up for global corporations to move in and take over.

Where Does This Leave the Gaian Worldview?

The vested interests of the powerful, justified and sustained by an economic ideology having little basis in reality, 'trickle down' to create a powerful vested interest the rest of us in the rich world share, to one degree or another.

Take-up of a Gaian worldview depends upon people seeing the interconnectedness all around them and joining the dots. This is the easy part; and there is evidence that it is already starting to happen. The real difficulty lies not in persuading people to listen to new ideas, but in persuading them to let go of old ones.

Neoliberal ideology has seeped into our collective consciousness: we are led to believe that material wealth is the most important measure of wellbeing; we are told so often that what we want is more 'choice' that we start to believe that choice is what we are actually being offered; and we are told that 'there is no alternative' to this system, when there are any number of alternatives yet to be tried.

For a Gaian worldview to gain purchase, there will need to be some major changes in the current system, starting with politics. Only governments have the power and legitimacy to stand in the way of the neoliberal juggernaut. But, as we have seen, 'politics as usual' is tied up with 'business as usual'. Only when we have properly democratic systems of governance that put people before corporations, and wellbeing before economic growth, will the Gaian worldview stand a chance of making a real difference.

Sources

American Heritage Dictionary of the English Language (2004), Fourth Edition, (Boston, MA: Houghton Mifflin Co.).

Gray, John (2004), *Al Qaeda and What It Means To Be Modern* (Faber and Faber).

Rees, William (2002), 'Globalization and Sustainability; Conflict or Convergence', *Bulletin of Science, Technology and Society*, **22** (4), pp. 249–268.

Wallerstein, Immanuel (1996), *Historical Capitalism* (London: Verso).

Richard Betts

Human-Caused Climate Change

The Evidence

These days we are surrounded by debate and discussion about climate change. It is a complex scientific problem which is still not completely understood, and its implications could have major consequences for the human species and indeed the rest of the world. Moreover, human actions to reduce climate change and adapt to its effects could also have major implications. Inevitably, then, the issue is the topic of heated debate. Opinions range from 'there is no problem', through 'there is a problem and we can address it' all the way to 'it is already to late to act'. In order to make informed decisions about our responses to the issue, we require robust scientific understanding of the issue and the likely consequences of our actions, or at least to have some grasp of the range of potential consequences if we are unable to be certain. This chapter reviews the latest scientific conclusions about recent climate change and its causes, and discusses the outlook for the coming century.

One of the iconic measures of climate change is the global average temperature near the surface. This can be established for roughly the last 150 years from a worldwide network of weather stations on land and observations made on board ships. In some places the observed temperature record extends further back, but before 1860 the worldwide coverage is not sufficiently dense to provide a credible global figure. The records show that global average temperature has risen by 0.74 degrees Celsius since the start of the 20th century. The rise has not been steady — before 1940 there was a warming of around 0.3°C, then there was a cooling of approximately 0.2°C until 1950, followed by a renewed warming of 0.13°C per decade since then.

Eleven of the last twelve years rank among the twelve warmest years on record, so the world has been warmer over the last decade than at any time in the observational record. This warming is observed over the oceans as well as over land, suggesting that it is a truly global phenomenon and not a conglomeration of 'local' warmings caused by some small-scale process such as the urban heat-island effect. The fourth Assessment Report (AR4) of the Intergovernmental Panel on Climate Change (IPCC), published in 2007, concluded that 'warming of the climate system is unequivocal'.

But what about before our global network of thermometers was established, or indeed before thermometers were even invented? 150 years is not long in the history of the Earth, and to establish whether the current warming is unusual we need to know more about temperatures further into the past. Temperatures can be estimated from a variety of 'proxy' evidence such as the patterns of growth in the rings of ancient trees, the distribution of particular species as indicated by their pollen found in the soil, and the chemical composition of air bubbles trapped in ancient ice. A number of independent studies have used these lines of evidence to reconstruct northern-hemisphere temperatures over the last one or two thousand years, and while they do not agree with each other perfectly, they all indicate that the warming measured over the last century is unusually rapid compared to the last two millennia.

The last century has also seen a rapid expansion of the human population and an even more rapid increase in our ability to modify the character of the Earth's surface and the chemical composition of its atmosphere, oceans and fresh water. It is well-established, and indeed not disputed by anybody with a scientific education or even basic common sense, that such changes have the potential to alter the Earth's climate. It is hardly necessary these days to provide an explanation of the well-known 'greenhouse effect', whereby certain 'greenhouse gases' absorb and re-emit some of the heat radiation given off by the Earth's surface and hence increase the temperature of the lower atmosphere. The most important greenhouse gas is water vapour, and without its warming presence in the atmosphere the Earth's average temperature would be around $-20°C$. Humans are not appreciably altering the amount of water vapour in the atmosphere, at least not directly, but we are increasing the amount of some of the other greenhouses gases such as carbon dioxide, methane and nitrous oxide. We have also introduced new greenhouse gases of our own, the 'halocarbons' such as chlorofluorocarbons

(CFCs) which have also damaged the ozone layer in the strato-sphere. While carbon dioxide, methane, nitrous oxide and the CFCs by themselves do not contribute as much to the overall greenhouse effect as water vapour, the fact that they are increasing nevertheless leads to an enhancement of the greenhouse effect and hence a warm-ing influence on climate.

Further records of air bubbles in ancient ice show us that carbon dioxide is now at its highest concentration for more than 650,000 years, and other records suggest that it may actually be higher than any time in the last several million years. We know for certain that combustion produces carbon dioxide, and we know for certain that we have been burning fossil fuels such as coal, oil and natural gas at ever-increasing rates over the last 200–300 years. From international records of energy consumption, is easy to calculate the quantity of carbon dioxide that has been produced by burning fossil fuels, and it is more than enough to account for the rise in carbon dioxide in the atmosphere. Moreover, we have also been clearing the world's for-ests to make way for farmland, and since forests lock up carbon within their biomass, their removal (which again often involves combustion) inevitably leads to a further release of carbon dioxide to the atmosphere. Therefore we are certain that the observed increase in carbon dioxide in the atmosphere has resulted from a combina-tion of burning fossil fuels and deforestation. It is less easy to deter-mine the exact contribution of each of these sources, but the current best estimate is that fossil–fuel burning has contributed approxi-mately three-quarters of the current excess of carbon dioxide above pre-industrial levels, and deforestation has provided approximately the remaining quarter. Cement production also makes a small con-tribution.

Methane and nitrous oxide, two other greenhouse gases, are also at record high levels. Most methane emissions and one-third of nitrous oxide are from human activities, largely agriculture.

At the same time as increasing greenhouse gas concentrations, we are also increasing the already-vast number of particles in the atmo-sphere. These particles are technically known as 'aerosols', but are not to be confused with the spray-cans of the same name. As with greenhouse gases, some kinds of aerosols occur naturally, examples being dust, volcanic ash, sea salt, spora from plants and dimethyl-sulphide ('DMS') from plankton. However, many are produced by human activity, mostly from burning fossil fuels. Burning wood also produces aerosols, so forest clearance and the use of fuelwood both increase aerosol concentrations. Additionally, desertification can increase the release of dust.

Aerosols have complex effects on climate. One effect is a cooling, since many of the particles cause some of the sun's energy to be reflected back into space, either directly through their own brightness or indirectly by increasing the brightness or lifetime of clouds. But aerosols can also absorb some of the sun's energy from that given off by the Earth's surface, with both processes giving rise to a warming effect. The overall effect of all aerosols is a cooling, although the precise strength of this cooling is less certain than the warming effect of greenhouse gases since it is more difficult to measure.

As well as affecting climate through increasing the concentrations of greenhouse gases and aerosol particles, changes in forest cover can also affect climate directly by altering the physical character of the Earth's surface. In particular, the proportion of the sun's radiation reflected back to space depends on whether the surface is bright or dark. A dark forested landscape absorbs most of the incoming solar radiation keeping itself relatively warm, whereas a darker unforested landscape reflects more radiation away which exerts a cooling effect. This effect is accentuated when the underlying surface is bright, such as in snowy conditions; anyone flying in a aircraft across northern Europe or Canada in winter will see that the forests stand out as dark areas contrasting with the brighter, snow-covered clearings and tundra.

In contrast, forests in warmer, moister regions can exert a cooling effect by promoting evaporation. Trees can capture a significant proportion of rainfall on their leaves and allow it to re-evaporate back to the atmosphere. Moreover, the deep roots of trees allow them to access moisture in the soil at much greater depths than grasses, so draw up more water through their trunks and branches into their leaves, from which it evaporates (or 'transpires') through microscopic pores into the air. Since evaporation exerts a cooling effect (think how cold you get when you are wet!), reduced evaporation / transpiration caused by tropical deforestation can lead to a warming. Furthermore, less evaporation can mean less water in the atmosphere to form clouds, so large-scale tropical deforestation could potentially lead to reduced cloud cover and an additional warming effect.

Since an unusual rise in temperatures across the globe has coincided with a unique and man-made rise in the concentration of gases known to exert a warming influence on the Earth, this suggests that the observed global warming is at least partly due to these greenhouse gas rises and that it is therefore at least partly a consequence of human activity. However, while this provides strong circumstantial evidence for a human influence on climate, it does not provide a rigorous scientific test of the theory. There are many other processes which can cause climate change — for example, changes in the output of energy from the sun, or changes in the Earth's orbit or the tilt of its axis which affect how much of the sun's is received by the Earth and the distribution of this across the Earth's surface. Large volcanic eruptions can inject very large quantities of aerosol high into the atmosphere, where they can spread around the globe and cool the Earth by blocking solar radiation. Also, as well as these 'externally-forced' variations in the energy received from the sun, natural 'internal' variability in the climate such as shifts in ocean currents and wind patterns can lead to warmer and cooler periods over years, decades and longer. To be confident in the causes of the current warming, and hence make predictions about the future, it is necessary to go beyond mere correlations and do more rigorous scientific studies.

The established scientific method for explaining a phenomenon is to carry out a controlled experiment, in which two (or more) samples are examined, with one sample being subject to a deliberate change while another is held unchanged. Clearly, this method cannot be applied to the Earth, since we only have one! However, it is possible to construct a 'virtual Earth' using well-established laws of physics and measured chemical and biological processes, and conduct controlled experiments on this instead. Such a 'virtual Earth' takes the form of a computer model, brings together a vast array of understanding of how the Earth's atmosphere, oceans and life work, and represents the processes in the form of mathematical equations solved by a computer program. The equations themselves have been established over the last few hundred years from careful observations, measurements and experiments by thousands of scientists from Isaac Newton onwards, both in laboratories and in the outside world. They are central to other aspects of physics, chemistry and biology as well as being the building blocks of computer models of climate. The climate models bring these equations together to provide an integrated view of the workings of the planet, and once again the models are tested and refined by comparison against observations. The very same computer models are used to provide weather forecasts on a daily basis, and the fact that such models are now able to provide accurate weather forecasts lends confidence to the idea that they are reasonably good representations of how the world works.

With our 'virtual Earth', we can now 'play God' and subject our mathematical planet to changes such as increases in the concentrations of greenhouse gases and aerosols, changes in forest cover and changes in the energy received from the sun. We can examine the effects of these acting together and in isolation from each other. We can also examine a climate state with no external influences, to assess the magnitude and rate of 'internal variability'. By comparing a variety of such simulations with the observed record of past climate, and seeing which model set-up agrees best with reality, we can establish further evidence for the causes of climate change. This technique is more sophisticated than simply comparing year-by-year global average temperatures; the geographical patterns of change are also compared, allowing the 'climate fingerprint' of different causes of change to be examined against the 'fingerprint' of the real change, and climate change is only attributed to a particular cause (or set of causes) if the fingerprints agree within established bounds of statistical significance. While this technique obviously relies on the model being realistic, it should be remembered that the models are grounded in well-established science and are tested against other data.

Such 'detection and attribution' studies have been carried out by a number of independent groups of climate scientists around the world, and all agree that the rise in temperature observed over the last 30-40 years cannot be explained without the rise in greenhouse gas concentrations. If only natural factors are taken into account, the computer models do not produce a warming over this period. More crucially, the change in the energy received by the sun over this period has not increased, so this cannot be the cause of the warming. Neither has there been any reduction in volcanic activity which might have contributed to warming. Internal variability in the climate system does not appear to produce such rates of warming. The conclusion, therefore, is that the observed warming is due to the increasing concentrations of greenhouse gases, and is therefore due to human activity.

Somewhat ironically, the full rate of warming has not been realised because of the corresponding increase in aerosol concentrations which is exerting a cooling effect to partly offset the warming by greenhouse gases. Moreover, the rate of rise of carbon dioxide in the atmosphere is only about half of the rate of emissions from fossil fuel burning, because some of the carbon dioxide is being absorbed by the world's vegetation and ocean waters. Therefore, we have been buffered from the full effect of our greenhouse gas emissions, partly by a service provided by the biosphere, and partly by a further consequence of our own pollution.

The implication of this is that continued burning of fossil fuels will inevitably lead to a further warming of climate. The complexity of the climate system is such that the extent of such warming is difficult to predict, but the same computer models that are used to attribute climate change to its causes can be used to provide an estimate of future warming if provided with scenarios of greenhouse gas and aerosol emissions.

The models assessed by the Intergovernmental Panel on Climate Change project further warming of between 1.0°C and 6.4°C by the end of the twenty-first century. Sea level is projected to rise by between 28cm and 58cm, and snow and ice are projected to shrink. Some models suggest that the Arctic could be ice-free in late summer by the latter part of the 21st century. Heat waves and extreme rainfall events are projected to increase. The intensity of tropical cyclones is projected to increase.

David M Wilkinson

Do We Need To Worry About the Conservation of Micro-Organisms?

One of the themes that marked out the Gaian approach from early in its development was an emphasis on the importance of microorganisms in the functioning of the Earth System. One reason for this was the involvement of Lynn Margulis, a distinguished American microbiologist, as a co-author with James Lovelock on several of the early scientific publications on Gaia. This emphasis on microbes contrasts with the more conventional approaches to studying ecology. Look in university level ecology textbooks or the academic journals and most of the organisms discussed will be large enough to see without a microscope. Yet for most of the history of life on Earth the only organisms were microbial, and microbes are key to many of the fundamental ecological processes on our planet. For example much of the photosynthesis on Earth—and associated oxygen production—is carried out by marine microbes (the phytoplankton). Microbes are also crucial to the cycling of nitrogen and the breakdown of organic matter (which releases nutrients for use by other organisms) as well as being involved in many other Gaian processes (Wilkinson, 2006).

Although microorganisms have no formal definition in biology they are usually informally defined as any life form too small to be seen without the aid of a microscope. A cut-off point of 0.5mm is sometimes suggested, a size just visible as a speck if viewed against a plain contrasting background with the naked eye. As well as the traditional bacteria (often now split into two groups, the 'Bacteria' *sensu stricto* and the 'Archaea') this definition includes viruses, many

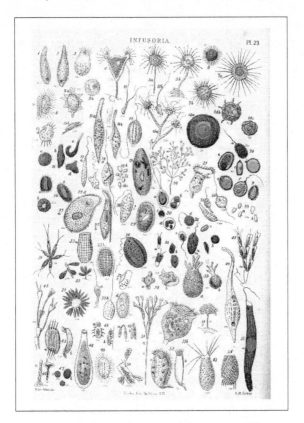

Figure 1. Hidden biodiversity invisible to the naked eye.

This plate is from a nineteenth-century text on micros-copy (Griffith and Henfrey, 1875) and illustrates an array of larger microorganisms. They range in size from being just visible to the unaided eye (about the size of the full stop at the end of this sentence), to requiring a magnification of around x400 to see even a small amount of detail. They are described under the archaic label of 'Infusoria' — a catch-all term for animal-like (i.e. they move when you watch them under the microscope) microbes. We would now classify most infusoria as Protists, or more informally protozoa. As well as being ecologically important many of these organisms are just as strikingly beauti-ful as the more familiar larger organisms which domi-nate the television nature programs and the concerns of most conservationists.

algae, protozoa and some fungi (Fig 1). Because viruses are parasitic on other cells, and the central concern of this chapter is free-living rather than parasitic microbes, they will not be considered in any detail here. This whole teeming world of life, invisible to the naked eye, only came to scientific attention during the second part of the seventeenth century with the invention of the first simple micro-scopes. However it was the end of the nineteenth century before the study of microbes in the natural environment (soil, water, etc.) really took off; this was particularly associated with the work of Martinus Beijerinck in Holland and Sergei Winogradsky in Russia. Indeed Russian 'ecology' has long had more Gaian tendencies than the ecol-ogy which developed in the West, emphasising two-way interac-tions between life and the nonliving aspects of the environment and the role of microbes in what, at first glance, may appear to be purely geological processes (Lekevièius, 2006). Some of the questions about the diversity of microbes which these pioneering microbial ecolo-gists addressed over 100 years ago are still controversial today — and have formed the subject of some of my own research since the mid 1980s (e.g. Smith and Wilkinson, 1987) and the subject of this chapter.

As microbes are central to most important ecological processes it raises the question; 'should they be of interest to nature conserva-tion?' The answer to this question depends, in part, on the likelihood of microbial species becoming extinct. As a by-product of their usual concerns conservationists are already involved in protecting the habitat of symbiotic microorganisms; that is microbes which live on or in other organisms in a harmful, beneficial or neutral manner. Most large organisms which have been studied in detail have para-sites and other symbionts living in them which appear to be restricted to that species — or sometimes a group of closely related species. For example the micro-fungi which cause powdery mildew in grasses have a range of specialist forms which are each restricted to one or two species of grass (Ingram and Robertson, 1999). So when a large organism becomes extinct, several species of microbes and other parasites (such as ticks, fleas and various 'worms') may go extinct as well. Therefore conserving an animal such as the Giant Panda is likely to have the effect of conserving associated microbes as well. Indeed, as well as conserving these organisms, conserving the Panda in the wild conserves the habitat for countless soil-living and other microbes (Fig 2). This is the flagship species approach to nature conservation; where the protection of a charismatic 'flagship'

Figure 2. The summit of Green Mountain on Ascension Island in the tropical South Atlantic.

The plants are obvious in the photograph, but there is also a hidden microbial biodiversity. For example examination of a single sample of soil from under the bamboo (surrounding the pond in the picture) produced eight species of protozoa, while 13 species were found from a sample of moss growing on a bamboo stem (Wilkinson and Smith, 2006). All of these species have wide geographical ranges and habitat tolerances, also being found in polar habitats as well as in this tropical cloud forest! The diversity of microbes is so poorly understood that before this work, by Humphrey Smith and myself, nothing was known of the free-living protozoa on the island. Ascension is not unique in this — many parts of the world have almost nothing known about the natural history of their microorganisms, and yet these microbes fulfil crucial roles in the functioning of the Gaian system!

species is used to conserve a whole range of equally interesting, if less telegenic, organisms which share its habitat (Simberloff, 1998).

A more complex question is, should we worry about the free living microbes, in soil and water, which are involved in so many ecological processes? A long standing maxim in microbiology has been 'Everything is everywhere, but, the environment selects'. It appears to have been first cited in this form by Lourens Baas Becking in 1934; although very similar ideas were suggested by Beijerinck and other late nineteenth-century microbiologists (de Wit and Bouvier, 2006). The idea is that because microbes are both very small and very numerous they are able to find their way to anywhere on Earth.

Their small size means that they are easily dispersed by air movements and their colossal numbers mean that even very unlikely events are going to be common in the microbial world. This can be understood by an analogy with lotteries; if you have enough lottery tickets you are highly likely to win, although the chance of any particular ticket winning is tiny. The unlikely event (such as being transported across the Atlantic in mud on bird's feet) is going to happen to some 'lucky' individuals of a particular species of microbe purely because there are so many of them. However this argument appears to run counter to what the microbial ecologist actually sees in nature. Take a sample of soil or water from somewhere on Earth and you won't find all species of microbe actively living in it. Martinus Beijerinck and his successors explained this apparent anomaly by pointing out that the correct habitat for a particular species of microbe will not be present at all locations. However the global distribution of microbes means that they will be found at all locations where the correct habitat exists. So, expanding Baas Becking's famous maxim; every thing is everywhere, but, the environment selects the species which can successfully grow and reproduce at a particular location. This is very different to what we are used to for larger organisms. For example suitable habitat for Prickly Pear Cactus was available in the Mediterranean parts of Europe and Australia for tens of thousands of years but until humans solved its distribution problem it was stuck in the Americas. Now it is an invasive pest in many semi-arid parts of the world.

If this idea is correct then there is little danger of free-living microbial species becoming extinct — for example even if you wiped out a species in equatorial Africa it could recolonise from tropical South America. However, there is currently no consensus in the microbial ecology literature on the correctness of this idea. While there is general agreement that microbes tend to be more widely distributed than many larger organisms the extent to which this is the case is contentious — see Finlay (2002) and Foissner (2006) for rival views of this question.

Some free-living microorganisms appear to be undoubtedly not 'everywhere'. A classic example is *Nebela vas*, a species of shell forming protozoa. This is found widely in mossy and organic rich soils, mainly in the southern hemisphere. The interesting thing about this species is that its shell has a distinctive shape (so it's unusually easy to identify under the microscope) and that it is absent from the parts of the world which have the longest history of microbial ecological studies (such as Europe and eastern North America) – so it is missing from exactly those areas where people have looked the hardest for such organisms. Soils rich in moss and organic matter are common in many of the parts of the northern hemisphere from which this species is absent, so in this case a large microbial species (around 0.15mm in length) has not dispersed to all suitable habitats (Smith and Wilkinson, in press). In other work, comparing the distributions of a particular group of protozoa in the Arctic and Antarctic, I have suggested that such relatively large microbes are often not found everywhere and so run counter to Baas Becking's famous maxim; however protozoa smaller than 0.10mm did tend to be found in similar polar habitats in both hemispheres (Wilkinson, 2001).

Bacteria are substantially smaller than these protozoa; for example several million can easily be found on a square cm of human skin, at least on the sweatier parts of our bodies. This statistic illustrates the colossal extent of unseen microbial diversity, albeit in a way that many people would prefer not to think about! If we apply the logic of Beijerinck and Baas Becking such extremely small organisms should be everywhere, indeed in recent years Bland Finlay and his co-workers have strongly argued that this is the case (e.g. Finlay 2002). In the past one of the problems in trying to study the distribution of bacteria in nature has been that they are morphologically rather simple organisms which often tend to look very similar to each other under the microscope. Recently the techniques of molecular biology have started to be used in such studies – effectively looking for distinctive 'genes' in the samples rather than whole organisms. These approaches have been suggesting that classical methods (i.e. trying to isolate and grow microbes from soil samples, etc, under laboratory conditions and examining them under microscopes) have grossly underestimated the number of bacterial species. For example one recent study looked at bacteria on the leaves of three species of trees in a Brazilian tropical forest, using DNA fingerprinting techniques similar to those well known from countless books and TV programmes on forensic science (Lambais et al., 2006). These scientists estimated that their results implied that there were several hundred bacterial species on these leaves, almost all of which were unknown to science. Such results are common when these 'molecular' techniques are applied to identify bacteria in soil and water samples (Hughes Martiny et al., 2006). However another recent study of soil bacteria across North and South America found limited evidence for such local bacterial 'species' and suggested that the main factor controlling which types of bacteria were found at any given sampling location were aspects of the soil chemistry (Fierer and Jackson, 2006). This study exactly fits the expectations of the 'everything is everywhere, but, the environment selects' maxim of early twentieth century microbiology. The key point about these contrasting recent studies is that their contradictory conclusions serve to highlight the fact that we are still some way from understanding the distribution and diversity of free living bacteria, even with the use of the powerful new methods of molecular biology. One of the problems with interpreting such results is that it is not clear what ecological interpretation should be put on the genetic differences identified in the DNA of free living microbes. For example, how different in their genetic make up should bacteria be before we consider them different species? And how ecologically distinct are such bacterial species? In addition if many of these 'species' do the same ecological 'job' (such as breaking down soil organic carbon), then it *may* not matter to the functioning of ecosystems if some become extinct. But we could do with being sure about this!

There is still much we don't know about the distribution and diversity of free-living microbes. Given their importance, rectifying this must be a major objective for twenty-first-century ecological research. However, it appears at least possible that human-caused extinctions of free living microorganisms are likely to be a lesser problem than is the case for larger organisms because the microbes often have large population sizes and wide distributions (even if the distributions turn out not to be as global as some scientists currently suggest). Given the crucial role microbes play in the working of Gaia this is an encouraging conclusion—although we urgently need to develop our understanding of microbial ecology to the point where we can have real confidence in these conclusions. However, the widespread extinctions of larger organisms must be causing a corresponding destruction amongst symbiotic microbes living on or in them. Indeed many of these microbes will also harbour parasites themselves—for example bacteriophages (viruses which infect bacteria)—so the loss of biodiversity caused by what appears to be a single extinction of some mammal, bird, insect or plant could be substantially greater than it appears to a conservationist without a microscope. Every time we read of a species becoming extinct, we should also lament the loss of a fascinating collection of microbes whose habitat was provided by that now extinct larger organism and also worry about the effects of all extinctions on the functioning of Gaia, be they of large or small organisms.

Sources

Note: Most of these references are to publications describing original scientific research in rather technical terms — written for a scientific readership. See Wakeford (2001) for a highly readable 'popular' book on the importance of microbes.

de Wit, R and Bouvier, T. (2006), *'Everything is everywhere, but, the environment selects*: what did Bass Becking and Beijerinck really say?', *Environmental Microbiology*, **8** (4), pp. 755–758.

Fierer, N. and Jackson, R.B. (2006), 'The diversity and biogeography of soil bacterial communities', *Proceedings of the National Academy of Sciences (USA)*, **103** (3), pp. 626–631.

Finlay, B.J. (2002), 'Global dispersal of free-living microbial eukaryote species', *Science*, **296**, pp. 1061–1063.

Foissner, W. (2006), 'Biogeography and dispersal of micro-organisms: a review emphasizing protists', *Acta Protozoologica*, **45**, pp. 111–136.

Griffith, J.W. and Henfrey, A. (1875), *The Micrographic Dictionary* 3rd ed. (London: John Van Voorst).

Ingram, D. and Robertson, N. (1999), *Plant Disease* (London: Harper Collins).

Hughes Martiny J.B. and 15 others (2006), 'Microbial biogeography: Putting microorganisms on the map', *Nature Reviews Microbiology*, **4**, pp. 102–112.

Lambais, M.R., Crowley, D.E., Cury, J.C., Büll, R.C. and Rodrigues, R.R. (2006), 'Bacterial diversity in tree canopies of the Atlantic forest', *Science*, **312**, p. 1917.

Lekevièius, E. (2006), 'The Russian paradigm in ecology and evolutionary biology: *pro et contra*', *Acta Zoologica Lituanica*, **16** (1), pp. 3–19.

Simberloff, D. (1998), 'Flagships, umbrellas, and keystones: Is single species management passé in the landscape era?', *Biological Conservation*, **83**, pp. 247–257.

Smith, H.G. and Wilkinson, D.M. (1987), 'Biogeography of testate rhizopods in the southern temperate and Antarctic zones', *Colloque Sur Les Ecosystems Terrestes Subantarctic, CNFRA*, **58**, pp. 83–96.

Smith H.G. and Wilkinson, D.M. (in press), 'Not all free-living micro-organisms have cosmopolitan distributions: The case of *Nebela (Apodera) vas* Certes (Protozoa: Amoebozoa: Arcellinida)', *Journal of Biogeography*.

Wakeford, T. (2001), *Liaisons of Life* (New York: John Wiley).

Wilkinson, D.M. (2001) 'What is the upper size limit for cosmopolitan distribution in free-living microorganisms?', *Journal of Biogeography*, **28**, pp. 285–291.

Wilkinson, D.M. (2006), *Fundamental Processes in Ecology: An Earth Systems Approach* (Oxford: Oxford University Press).

Wilkinson, D.M. and Smith, H.G. (2006), 'An initial account of the terrestrial protozoa of Ascension Island', *Acta Protozoologica*, **45**, pp. 407–413.

Patricia Spallone

The Gaia Effect

Making the Links

'Gaia … is difficult to describe', reflects James Lovelock.

> The nearest I can reach is to call Gaia the theory of an evolving
> system — a system made from the living organisms of the Earth,
> and from their material environment, the two parts being tightly
> coupled and indivisible … this close coupling of organisms and
> their environment is strong enough to have greatly influenced
> the way in which the life-environment system on Earth, and on
> other planets with life, has evolved. (Lovelock, 1990, pp. 100–101)

But let's be clear. 'My Gaia', the microbiologist Lynn Margulis adds,

> is no vague, quaint notion of a mother Earth who nurtures us.
> The Gaia hypothesis is science. The surface of the planet, Gaia
> theory posits, behaves as a phsyiological system in certain lim-
> ited ways. The aspects that are physiologically controlled
> include surface temperature, atmospheric composition of reac-
> tive gases, including oxygen, and pH or acidity-alkalinity.
> (Margulis, 1998, p.123)

It is not difficult to see how Gaia theory has inspired the environmental movement. But it has also inspired other social thought and
action, and has offered a paradigm to academic studies in politics,
economics, theology, development studies, urban design, and more.
James Lovelock calls Gaia a protean idea; an idea that can take on different forms. (Lovelock, 2001) Gaia has thus been a springboard and
a model for thinking and action in many domains, not just within the
natural and physical sciences. An insight from the historian of science Ludmilla Jordanova is helpful here. Scientific ideas can be
understood as mediations, 'they speak to and contain implications
about matters beyond their explicit concern' (Jordanova, 1998, p. 42).

But there are also other, less explicit, links to be made with Gaian
thinking. In my own area of work on the social and political aspects

of biological sciences and biotechnologies, I see it in several places: at grass roots level where women's groups and NGOs (Non-Governmental Organizations) have responded to developments such as *in vitro* fertilization, cloning, genetic modification; in the practical discipline of medical ethics where the idea of the patient as autonomous individual, disconnected, is fragile; in social and feminist theory; and in the biosciences, in the apparent turn away from reductionism in genomics.

Taken together, the examples I have in mind suggest a social transformation which I will call the Gaia effect. I am only just starting to think about the possibilities and why they are important. This is the subject of this chapter, which is a very personal account.

Making Links

In Bangladesh in 1989 an international women's conference on health and new reproductive technologies brought together activists, academics and other workers, to explore the range of experiences and priorities around health and reproduction, and in particular to explore novel developments in reproductive and genetic technologies (Spallone, 1992). We participants from the UK were looking most closely at assisted conception techniques and the expanding scope of medical genetic testing. Up to then, public debate had focused extremely on the moral status of the human embryo, the isolated human embryo of western intellectual thought. We wanted to raise questions that were not being raised in public forums. What do these new developments mean for women, for women's bodies where eggs and embryos come from? How does society view and support disabled people? Why these approaches? By contrast, several delegates from India and the Philippines arrived with a seemingly different main focus, genetic modification technology brought to agriculture and food production. Theirs was a cogent analysis about consequences for the environment, about risks to food security, to women's livelihoods, and to survival as a result of the control of the world's food resources by a handful of multinationals who were consolidating ownership of seed companies at the time, and developing a new generation of seeds with properties which no one had asked for (Spallone, 1993). They didn't make the conceptual distinctions I was between medical and agricultural technology and policy, between the human body and the land. The connections were self-evident to them, not just because everything affects everything, but because they were inseparable in their everyday lives. Such connections were not surprising. Other delegates from western countries presented on these issues. We had all been sharing analyses and concerns. But it nevertheless took me up short in that setting. These women came to a health conference to talk about agriculture, food and land in the same breath as we were talking about the body, medicine and reproduction, not as if it were an 'also relevant' theme.

Similar connections were being made on my doorstep, back home in the UK. Prompted by a draft directive of the European Commission (EC) allowing the patenting of life forms resulting from certain forms of gene and cell manipulation, the National Federation of Women's Institutes (WIs) tabled a resolution to be voted upon at their annual meeting in 1992: 'This meeting calls for the EC draft directive authorizing the patenting of life forms to be rejected until there has been more public debate concerning its legal, social, environmental and ethical implications.'

A highly informed background paper was circulated before the meeting which flagged, among other things, the impact on farmers and food growers in developing countries, who were feeling the earliest effects of a new range of genetically modified seeds and seed-herbicide packages. Would food growers there be obliged to pay royalties on the next generation of seeds to companies 'owning' and promoting the modified varieties to their governments? The EC directive was largely a response to novel agricultural biotechnologies. Many WIs are based in rural and farming communities, so it was not surprising that the issues and the technologies were important to them.

To further inform their discussions, the North Yorkshire East Federation and the Derbyshire Federation invited me to address three local meetings, in York, Scarborough, and Derby. These did not focus on patenting so much as to unearth other connections and questions. At one meeting, a member stood to say that she had recently read that a gene 'for' asthma was identified. At first she felt positive about this. Her son had asthma, you see. But then she wondered, would this mean that attention is diverted from dealing with air pollution and the environment?

This is one anecdote, but it would not surprise campaigners or social researchers who might put it this way: in everyday life, people make links with things related to each other. The distinctions we may make between, say, food and medical technologies can be help-

ful. We name and define problems for various reasons. But the isolation of these categories does not endure in other domains.

These two examples at 'ground level' suggest *a relational view of
human existence*, a phrase I have borrowed from the social anthropologist Jeanette Edwards (1993, p. 11). This is not far from a 'tight cou-
pling' of ourselves and the environment, a concept integral to Gaia
and which Anne Primavesi (2000) identifies as significant to other
areas of human life. There is a resonance between these two concepts
from two different domains of knowledge, and it has implications
for addressing environmental concerns (among others) such as the
one with which Gaia scientists are most engaged, climate change
and global warming. My point is that Gaia thinking is not out on a
limb. Many persons are thinking in the same register even if they
aren't calling their concerns 'ecological' or 'Gaian'. Recognising this
and building a bridge between the two seem to me to be crucial in
forwarding new environmental practices.

Medical Ethics and Bioethics

A shift in the same direction—toward a relational view of human
existence—can be found in the practical disciplines of medical ethics
and bioethics where the legitimacy of the principle of individual
autonomy and the model of informed consent on which it rests is
being challenged. In clinical research trials that require human participants, a most effective challenge has come from voices in the
developing world, where the idea of an isolated individual has been
so clearly alien (Nuffield Report, 2002). In this context, bioethics is
moving away from the notion of autonomous individuality that has
dominated the field, and moving toward a sense of relatedness as a
condition of moral evaluation and ethical decision-making. I am not
suggesting a clear-cut shift; individual informed consent remains an
important practice, and can be depended upon in new quarters of
genomics research, as one anthropological study of the Swedish
'biobank' experiment shows (Hoeyer, 2005). Yet, the fragility of an
individualised informed consent policy was apparent even here.
The research participants in this study, those involved as donors and
potential donors of blood and tissue for DNA and other biological
analyses, resisted identifying themselves as lone individuals. They
saw themselves as socially bounded. What makes this all the more
interesting is that genomics itself, exemplified by medical biobanks,
is demanding a sense of relatedness. Biobank research involves a
sense of individuals and families but also other groups, geographi-

cally bound communities, and national populations connected by virtue of a presumed shared biogenetic heritage or possible connections yet to be established.

This shift in areas of ethics—tentative, precisely located—is sympathetic to social anthropologists' understanding of relatedness, of a relational view of existence. I also see these shifts in ethics and in genomics as complementary to the turn away from reductionism found in the natural sciences, as Brian Goodwin describes in this book. I am calling it the Gaia effect. What I mean by this is that connections and relationships are in the foreground of thought and understanding now, and we are seeing them—they are not entirely novel to human experience or the sciences—with fresh eyes. This is an observation, but also something we can work with and build on: a starting point for renewed thinking and action on the environment, health, ethics and related things.

There are other links to be made and others to be found. Let me offer one more. To regenerate 'an ecological self' in the industrial world, the anthropologist Frédérique Appfel-Marglin and Suzanne Simon (1994) look to what we can learn from Oriya people of India. Here they find a conception of the world that makes no radical division between the human and non-human. The human and non-human are indivisible, you could say, along the lines of Lovelock's explanation of the relation between living organisms more generally and the environment. I am not doing the example justice here, but its resonance with a Gaia worldview taken to the realm of human consciousness and social life is striking. Further, it offers a big challenge. For what these researchers find among the Oriya is that there is no stark division between reproductive and productive behaviour. Rather, 'the body, the land and the community are all embedded in the same processes of generation and regeneration that necessitate the work of both men and women' (p. 41). Moreover, the bond with the earth is substantive. 'Human welfare and the welfare of the land are related by the consubstantiality [sharing of substance] between them' (p. 42). Consubstantiality is a rich concept that could do important work for us in forging a new ecological consciousness and a real way of being in the world.

Why Gaia?

There are sympathies, resonances, and contrasts among the examples in this account. In considering the terms of our relationship to the planet and ourselves, how might we manage to relate these various ways of thinking when we use them together in our lives? (Midgley, 2003). We can revisit what we thought we understood. We know some things, need to learn more, and have resources to draw upon. How are we to understand the proper relationship between human beings and the non-human world? Gaia theory talks about tight couplings. But Gaia scientific theory is not human-centred (nor should it be), and I would not wish to lose sight of justice and distribution of resources as part of the ecological project. This inclusion need not dilute the ecological project but rather should promote it.

Think about our response to climate change and global warming. Responses are needed at the levels of both governance and everyday life. Both are necessary to shaping the morality of the world we want to make. Both were critical for the Black civil rights movement in the US, for the women's movement, for disability rights movements. It is true for addressing global warming. Governance, the everyday and the domestic are elements of the ecological project of Gaia, which is a social project, a political project, a cultural project. There is a perfectly useful concept to hand in praxis, bringing together theory and practice. Crucial as ever is openness and criticism. What versions of the terms and concepts of biological theory are we using? What are the priorities of local people and communities, for example those who may be struggling with crime and poverty as their first concerns? (Wakeford et al., 2007)

But why posit a Gaia effect? You don't necessarily need it to define practices to address, say, inequalities in access to the world's food and energy resources. I am using Gaia as an idea, but not as though Gaia theory provides a factual biological basis for decision-making. I am caught in my own history here, in experience of the supposed biological basis of women's behaviour, of intelligence and race, of criminal tendency, of uses of 'nature' to naturalise ideas about human difference. So, I am not suggesting a linear application of scientific theory to explain social life (to do so would, in any case, not be science but scientism).

Gaia as an idea is perhaps best understood as a symbol of relationships among many real forces: forces in nature, political forces, forces that make us human beings. Things that we argue over. As Rebecca West put it, ideas are not 'things to pick and choose ... ideas are symbols of relationships among real forces that make people late for breakfast, that take away their breakfast, that make them beat each other across the breakfast table' (West, 1992, pp. 1085–6).

In this sense, Gaia is a plural idea. It is (or should be) open. It can be inspired by other ideas and practices from other places. Gaia is applicable now. It doesn't represent nature as prior to culture, which is to say it is not scientistic. It is transformative in that it demands action and can help define action. It allows us to mobilize other concepts and ideas to do this. It gives an opening to define the social values underlying our technology assessments. How might it do all this? By giving us an orientation to the world of which we are a part, an orientation that can take on board scientific explanations of things as well as social and aesthetic explanations of things; and to develop an ecological praxis without losing sight of each other, what the anthropologist Margaret Lock (2003) in another context called the 3Ps: persons, politics and poverty. Or perhaps the power of the idea of Gaia is simply that it can at this critical moment hold together all the pieces into a single vision.

In the end, the answer to 'why Gaia?' is very personal. Gaian perspectives have stimulated me to think in another register, to be open to another way of thinking. Earthbound, it doesn't hold a grudge against a sense of the spiritual, which may or may not have its source in a religious tradition. Umberto Eco (2002) finds a 'lay ethics' in sensitivity to our connectedness to the 'cosmic substance'. The scientist of complexity Stuart Kauffman (1995) looks for it not in mysticism but in mathematical laws of the universe. At the same time, it allows that there are real conditions of living though many ways of knowing them.

Sources

Appfel-Marglin, F. and Simon, S.L. (1994), 'Feminist Orientalism and Development', in *Feminist Perspectives on Sustainable Development*, ed. W. Harcourt (London: Zed).

Eco, U. (2002), `When the Other Appears on the Scene', in *Five Moral Pieces* (London: Vintage).

Edwards, J. (1993), 'Explicit connections: ethnographic enquiry in north-west England', in: *Technologies of Procreation: Kinship in the Age of Assisted Conception*, J. Edwards, S. Franklin, E. Hirsch, F. Price, M. Strathern (Manchester University Press).

Hoeyer, K. (2005), 'Studying ethics as policy: the naming and framing of moral problems in genetic research', *Current Anthropology*, 46 (Supplement), pp. S71-S91.

Jordanova, L. (1989), *Sexual Visions: Images of Gender in Science and Medicine between the Eighteenth and Twentieth Centuries* (New York: Harvester Wheatsheaf).

Kauffman, S. (1995), *At Home in the Universe: The Search for Laws of Self-Organization and Complexity* (Oxford: Oxford University Press).

Lock, M. (2003), 'The Hype of Genomics and the Mirage of Health', Plenary lecture, Vital Politics Conference, BIOS Center, London School of Economics (September).

Lovelock, J. (2001), *Homage to Gaia: The Life of An Independent Scientist* (Oxford: Oxford University Press; first published in 2000).

Lovelock, J. (1990), 'Hands up for the Gaia Hypothesis', *Nature*, **344**, pp. 100–102.

Margulis, L. (1998), *Symbiotic Planet: A New View of Evolution* (New York: Basic Books).

Midgley, M. (2003), *The Myths We Live By* (London and New York: Routledge).

Nuffield Council on Bioethics (2002), *The Ethics of Research Related to Healthcare in Developing Countries* (London: Nuffield Council on Bioethics).

Primavesi, A. (2000), *Sacred Gaia: Holistic Theology and Earth System Science*, (London and New York: Routledge).

Spallone, P. (1992), *Generation Games: Genetic Engineering and the Future for Our Lives* (London: The Women's Press; Philadelphia: Temple University Press).

Spallone, P. (1993), 'The Emperor's New Genes', *Science as Culture*, **18**, pp. 110–133.

Wakeford, T. with Singh, J., Murtuja, B., Bryant, P. and Pimbert, M. (in press 2007), in *Handbook of Action Research* (Second Edition), eds. H. Bradbury and P. Reason (New York: Sage).

West, R. (1994), *Black Lamb and Grey Falcon: A Journey Through Yugoslavia* (London: Penguin Books; first published in 1941 by Viking Press).

Anne Primavesi

Can Gaia Forgive Us?

When the topic 'Can Gaia forgive us' was given me to debate at an interfaith meeting, some underlying questions had first to be addressed. If, as appears to be the case, the title question reveals a genuine awareness of the harm that we have inflicted on Gaia, what is our purpose in seeking forgiveness? Is it to assuage our guilt for what we've done or does it simply express the desire to escape from the consequences of our actions? Furthermore, if forgiveness is asked for by us but can only be granted by Gaia, who am I to grant myself the right to forgive in Gaia's name? How can I speak for any-one but myself?

Then I remembered what Rabbi Hillel had said:

If I am not for myself, who will be?
And if I am only for myself, what am I?
And if not now, when?

How does speaking for myself relate to speaking for others and ulti-mately, for Gaia? Does my ability to speak for myself affect my abil-ity to speak for them—and vice versa? The word Gaia, as used by James Lovelock, clearly signifies a state of affairs in which I am (and I believe this to be the case) 'tightly coupled' with a very complex global living system of interacting living and non-living compo-nents. For him and for other scientists this signifies that Gaia is an object of scientific enquiry (Primavesi 2000:2–21).

But when I ask whether or not, when I speak for myself, I am also speaking for or certainly from within that system, then the term Gaia functions rather differently. It means more than a scientific reality to be investigated. It points to a community of living and non-living beings whose sum is greater than its parts; one in which I and every being inescapably play a part in the life of the whole.

So Rabbi Hillel's questions remind me that while I must speak for myself, I must also be aware that what I say affects not only myself. Speaking for myself can, therefore, be an opportunity to speak for *'more than'* myself. And what might that 'more than' include? Might it not, potentially at least, extend to the whole of the more-than-human community of life on Earth (Abram 1996: ix)?

A recent report helps pinpoint what is in question here. It found that whereas in 2005 only 45% of people in Britain thought we should look after our own interests rather than those of the community, by 2006 53% of us think our quality of life is best improved by looking after individual interests. Moreover this response assumes 'the community' comprises human beings alone. Even so, the majority presume their interests are best served not by looking after those of the community but by being 'only for myself'.

But, asks Rabbi Hillel, if that is so, what am I? Where do I belong? What kind of creature emerges from, is related to or is supported by itself alone? What kind of person cares solely about herself or himself? Or indeed can afford to do so. For we know that we are not islands, but part of the mainland and part too of the surrounding oceans.

What part do we play there? It has long been assumed that the oceans are too vast for humanity to damage in any lasting way. But we now learn that, over time, our overdosing them with basic nutrients—the nitrogen, carbon, iron and phosphorus compounds that curl out of smokestacks, wash into the sea from fertilized lawns and cropland, seep out of septic tanks and gush from sewer pipes—has altered their composition by making them more hospitable to primitive organisms.

So at certain times of year, along stretches of coastline in Europe, Australia and the Americas, blooms of cyanobacteria turn the sea into a stinking, yellow-brown slush that stains fishing nets and coats them with a powdery residue that constricts fishermen's throats, blisters their skin and burns their eyes. These ancient life forms flourished 2.7 billion years ago and now, where they thrive and spread, more complex forms of ocean life, such as fish, corals and kelp on which other marine lifeforms and human communities depend, struggle to survive. Where this pattern is most pronounced scientists evoke a scenario of evolution running in reverse, returning us and a vital Gaian life support system to living in the state of primeval seas hundreds of millions of years ago.[1]

Can Gaia forgive us for this? What would forgiveness mean in this situation? While such scientifically based facts increase awareness of the harm we have done and are doing to Gaia, it is only in and

---

[1]   See *Poison in the Seas* at http://timethief.wordpress.com/2006/08/08/altered-oceans-a-5-part-series-text/.

through the process and pain of relating to other human beings that we learn what forgiveness for such harm might mean: or what it asks of us. We learn too that the measure of forgiveness we aspire to for ourselves is the measure of what we can hope for from others for ourselves. This yardstick is expressed most precisely in the petition for forgiveness in the Lord's Prayer. It puts the case to God in the following terms: 'Forgive us as indeed we forgive others' (Matt. 6.12). The hinge of this famous saying, as John Caputo points out, is 'as indeed', *hos kai*. Forgive us, or not, as indeed we forgive each other (Caputo: 1997: 226).

What does it mean then to forgive each other? Christians widely use the term forgiveness as if its meaning is univocal and self-evident. Yet the Greek New Testament word *aphiemi* is translated 'forgive' just 47 times but used (as in Matt. 6.12) 146 times. On other occasions it is translated by words such as leave, let alone, dismiss, let go, forsake: in all of these the central idea is 'to let go': usually to 'let go' of pain and hurt but also, and very importantly, of the desire to inflict pain or hurt in return. This not only expands the meaning of our word 'forgive'. It also signals that forgiveness is a process; indeed a profound, painful and often lengthy undertaking.

This process was powerfully exemplified in the response of Anthony Taylor's mother to his murder. She said she had to forgive her son's racist murderers because Jesus forgave those who murdered him. Her whole demeanour spoke of the difficulty, indeed agony involved in the decision to take on this process; and of her awareness that it would require letting go of any hatred she might bear towards them. It was also clear that she knew she was committing herself to a task that would absorb her energies for a long time; in a prolonged relationship with those young men and with the communities to which they belong. She has said too that she is willing to meet one of the murderers who expressed repentance for his crime and, if they wished, she would meet his parents also. That will certainly affect the quality and progress of her life.

But not only hers. All those she meets will be affected throughout this process. One mark of authentic forgiveness is the reestablishment or healing of community relationships. This does not necessarily require all becoming bosom friends. But it does mean leaving room for healing and for hope that the communities' members may live together again in more enduring and healthy relationships.

That process is being hastened for her by one of the murderers showing repentance; even though she did not wait to forgive until he repented. Forgiveness by those injured is independent of repentance on the part of those inflicting the injury; just as repentance too is independent of forgiveness. Mrs Taylor could simply have 'let the murderers go'; dismissed them (apparently) from her life and taken no account of whether or not they repented or of what happened to them in the future. So too, whether or not we repent, scientists warn that we may be 'let go' by Gaia; dismissed from the community of life on Earth.

So what role could repentance on our part play in Gaia's forgiveness? The analogy works positively for us if, as it should do, repentance involves an awakening of our collective consciousness to the harm we have done and are doing—and if it is accompanied by a resolve to work toward forgiveness in a positive sense by redressing the harm where and how we can. For while forgiveness is not dependent on repentance, and either can come first, the one tends to encourage the other. The Greek word for repentance, *metanoia*, means 'to change one's mind', 'to feel sorrow for or regret', 'to be of a new heart', 'to change one's view and purpose'. Again, the language speaks of a process, of a long-term commitment that involves changing attitudes and purposes, turning one's life around as well as feeling sorrow and regret for hurt inflicted (Caputo 2006: 143–6).

The point to stress is that while repentance is independent of forgiveness, the *purposes* of repentance and the *purposes* of forgiveness are the same. So asking 'Can Gaia forgive us?' does not mean waiting for an answer before we repent. It presumes 'changing one's view and purpose' in regard to Gaia: acknowledging injuries inflicted by us on her more-than-human community together with a steadfast resolve to heal them. This in turn requires taking proper responsibility for maintaining the life support systems essential for the wellbeing of the whole Gaian community.

The injuries we inflict on those systems have a dual aspect: the collective, or structural one of those attributable to us as a species and that resulting from our activities as individuals. Structurally, whether we like it or not, our lives are in service to prevailing capitalist economic systems oriented toward monetary growth that relies on massive overuse of the earth's resources. Supporting those systems is a view of the world as fragmented into competing, warring factions in dispute over possession of resources such as oil and minerals.

This increases the demand for personal and national security which in turn is used to justify the growth of the global military-industrial complex. Its stated purpose is 'defence' against attack. Its real (unstated) purpose is to increase our ability to destroy, in the name of human security, some, or in many cases, all forms of life: directly through weapons technology and indirectly through massive allocations of money and human resources for purely destructive purposes. The latest figure for annual subsidies to the: arms export trade in Britain is £851.91 million (Thomas 2006). The government's use of terms like 'security' and 'defence' to describe its purposes makes it possible for it to allocate inordinate amounts of their budget to weapons development and derisory amounts to social and environmental welfare. The proposed allocation in £billions for the development of Trident nuclear missiles, compared to that being spent on energy-saving programmes, is a case in point.

Our injurious impact on Gaia as individuals is now generally assessed as the size of our personal ecological footprint: the amount of planetary resources each of us consumes. In the 'developed' world this is closely bound up with the prevailing consumerist culture that, driven by the need to sell more and more products made from ever-dwindling resources, relies on increased consumption in pursuit of the chimera of ever-expanding economic growth. Repentance here would aim for and encourage a culture of 'self-limitation': satisfying our vital needs in moderation and in accordance with what each of us finds possible. The superfluous as opposed to the necessary character of many of our purchases would then become apparent and might act as an efficient personal monitor.

The question 'Can Gaia forgive us?' now appears rather differently. It assumes (rightly) that we have hurt, victimized, damaged the more-than-human community of life on Earth. If so, Gaia's purposes and ours will be served by a repentance that acknowledges the wrong done and effects change in our lives that will heal relationships with the members of that community. And by realizing that as we do so, we are healing wounds inflicted on and by ourselves.

Part of the problem is that in an increasingly urbanized world we appear to interact almost exclusively with other human communities and increasingly now, with human-made technologies. These are what David Abram calls 'the very structures of our civilized existence — the incessant drone of motors that shut out the voices of birds and of the winds; electric lights that eclipse not only the stars but the night itself' (Abram 1996: 21–8). The emergence of climate change, however, as a 'hot' topic, reminds us forcibly that we still *need* that which is other than, more than ourselves and our own creations in order to live well and happily. The real question then becomes: Can our children forgive us for temperature rises of 3°C to 5°C by the year 2100?

So, asks Rabbi Hillel: *If not now, when*? Now is the time for repentance: for learning about and acknowledging the harm we have done; for changing our views and lifestyles and our attitudes towards global, national and local issues. Presently they are perceived and dealt with routinely through financial 'clout', trade restrictions and military 'solutions' strongly supported by prevailing notions of our relating to others primarily as competitors for resources; as buyers and sellers of commodities; and so as potential or real enemies.

Given this state of affairs it is not surprising that the concepts of forgiveness and repentance emerge from a religious rather than economic or political understanding of relationships. Religion, derived from *religio*, is properly interpreted as our being linked and re-linked together just as ligaments knit our bones to our bodies and enable us to function as a whole. Genuine religion unites us, giving status and significance to all within the community of life on earth: a status and significance endorsed for some of us by belief in God but not dependent on that.

From this viewpoint repentance is a proper response to the harm done to that community because it is based on knowing ourselves and our activities to be intrinsically embedded within it. If I am only for myself, conscience tells me, I am nothing. True, the word repentance can be used (by those so inclined) to impose a mode of conduct on themselves and others with a view to earning us a reprieve from Gaia's wrath. But then it is no longer a genuinely religious concept. As such it requires us to change our views and purposes because this is the correct response to present circumstances: one that accords with the demands of justice imposed by our status and significance in the world. For repentance, like forgiveness, emerges out of an understanding of what it means to live for and through 'more than' ourselves. Of what it means to 'turn away from dead and deadly works' (Heb. 6.1).

Sources

Abram, D. (1996), *The Spell of the Sensuous: Perception and Language in a More-than-Human world* (New York: Pantheon Books).

Caputo, J.D. (1997), *The Prayers and Tears of Jacques Derrida* (Bloomington, IN: Indiana University Press).

Caputo, J.D. (2006), *The Weakness of God: A Theology of the Event* (Bloomington, IN: Indiana University Press).

Primavesi, A. (2000), *Sacred Gaia* (London and New York: Routledge).

Thomas, M. (2006), *As Used on the Famous Nelson Mandela: Underground Adventures in the Arms and Torture Trade* (London: Ebury Press).

John Mead

The Human Psyche & the Imminence of Climate Catastrophe

'Everything begins in mysticism and ends in politics.'
Charles Peguy

'Human kind cannot bear very much reality.'
T.S. Eliot

Climate catastrophe is now imminent. There is ample evidence that this is largely the result of human behaviour, and therefore of the human psyche, chiefly in the rich countries, in producing emissions of greenhouse gases, mainly CO_2. There is good reason to suppose that this catastrophe will be part—probably the major part—of a more general collapse of our civilisation. It may well be the end of human existence.

But the traditional visitor from Mars, listening to our daily news bulletins, would not conclude that we were really anxious about the issue or giving to it any particular priority. Our culture is not behaving as though it were genuinely concerned to safeguard the future of our children or the integrity and health of our planetary resources. Our cultural attitude is a dramatic example of the psychological phenomenon of denial. Much of this chapter examines this phenomenon in more detail.

In attempting to avert catastrophe we are near a tipping point. We may have reached it. Several experts fear we may have passed it. At all events there is now very little time to spare—at the most ten years. If we fail to avert it, the prospect for our children will not be a gentle or peaceful matter. It is likely to be accompanied by conflicts—for increasingly scarce resources—of the most vicious kind, such as we are already seeing in Africa.

In short, if the climate scientists are right, our civilisation is in crisis. We are confronted by a very terrifying prospect. If we carry on as we are southern France and Spain are likely to become uninhabitable. Already half a billion people live in regions prone to chronic drought, due to global warming. By 2025 that number is likely to have increased five fold. There are already 25 million environmental refugees. The IPCC predicts that by 2050 that number will have risen to 150 million. The conference of the meteorological office in Exeter in February 2005 gave us at most ten years in which to make the very radical reductions in CO_2 emissions needed if catastrophe is to be averted. As George Monbiot has said, 'We are not facing the end of holidays in Seville because Seville is too hot. We are facing the end of human existence.'

As a species, what are we doing about it? In practical terms, the answer is: 'nothing'. To avert catastrophe we must as a species reduce our CO_2 emissions by 90% by 2030, from 1990 levels. But so far from reducing them, we are rapidly increasing them. In terms of Gaia, the future of the Earth is now in the hands of the collective human psyche, and is heading for disaster.

Let us spell out what needs to be done. Assuming we have not left it too late, averting disaster is technically quite feasible. It would require immense changes and immense cooperation. But it would require no genuine hardship. No one would have to be cold or go hungry. What is lacking is the human motivation to make it happen.

The fundamental problem was well summarised by the economist Richard Douthwaite: 'One can see that cuts (in CO_2 emissions) of the size needed are going to hit so close to home that it is impossible to see politicians taking the initiative and making them: only tremendous public pressure or a desperate crisis will get them to act'. Those words were written in 1992. Fifteen years later they remain the best summary of our predicament today. Public pressure has certainly increased, but it is still very far from being adequate to the task. And the nearest we have had in the UK to anything like a desperate crisis was the 2003 heat wave.

What is needed is a mass movement on a huge scale, claiming a supreme priority, drawing its strength from a very high degree of spiritual and ethical motivation, and, as Al Gore would say, 'forging a new common purpose' by bonding people together in vision and action.

But it is our motivation that is defective. Anyone who sets out to think through the imminence of climate catastrophe, the magnitude of the vested interests to be overcome, and the brevity of the time still available, is certain to encounter anxiety, and at times horror, depression and despair. Such a task is emotionally extremely demanding. It cannot possibly be undertaken without mutual sharing and support, and the bonding thus established. But these are the very things that denial impedes. In the words of a leading eco-psychologist, Joanna Macey:

> The perils facing life on earth are so massive and unprecedented that they are hard to believe. The very danger signals that should rivet our attention, summon up the blood *and bond us in collective action* [my emphasis] tend to have the opposite effect. They make us want to pull down the blinds and busy ourselves with other things.

In other words, the courage, enterprise, resolution, and above all the bonding in mutual support and cooperation — all the qualities needed for a mass movement — are secretly undermined by denial, by our refusal to recognise and share our feelings of anger, horror and despair.

As a result the vast majority of people simply do not think through the predicament. So they put no pressure on the politicians. Instead they themselves go into denial and apathy. That includes the highly educated and intelligent, and even those with a special interest in Gaia theory. They have plenty of other urgent and more rewarding things to do which do not involve them in horror and despair.

Just as climate catastrophe is beyond all question the greatest threat ever to confront humanity, so it should call forth by far the greatest common concern, effort, energy, and spiritual motivation. At the moment that is simply not happening.

Denial about climate change remains widespread in the UK public. But this is but one manifestation of a much more fundamental denial. Herman Daly in *Beyond Growth*, writing of the concept of sustainable development, refers to the shift it requires in our vision of how the economic activities of human beings are related to the natural world, one which involves replacing the economic norm of quantitative expansion (growth) with that of qualitative improvement. In his own words:

> This shift is resisted by most economic and political institutions *Enormous forces of denial are aligned against it*, and to overcome them requires a deep philosophical clarification, even religious renewal [my emphasis].

As the World Council of Churches has recently said: 'Measures to reduce greenhouse gas emissions run against the dynamic of the present project of society based in ever-expanding production and consumption. A vision of society is at stake'. Hence the need for denial to protect that vision. But what is meant by denial in this context? Obviously it is not just conscious deliberate dishonesty. The academic economists do not begin each day with the conscious resolve to tell lies to their students. Nor is it just ignorance, though ignorance certainly abounds. It refers rather to the honest rejection at a conscious level of some truth or fact which at a deeper level is known about in a confused way, but which is avoided because of the anxiety which it arouses. Denial is much more intractable than ignorance. The latter can be corrected by more information. But for denial, information is water off a duck's back.

Denial is a 'defence mechanism'. It defends the individual from some truth which he cannot afford to acknowledge because to do so would expose him to feelings of confusion or horror or shame. Denial is in fact the process by which, for the most part unconsciously, we manage or keep at bay what would otherwise be very disturbing emotions. It therefore defends that basic clarity and peace of mind which we all need if we are to carry on with our lives, and to which we all therefore tend to cling. To remove it plunges us into a struggle somehow to bring new order into what has become frightening disorder in our understanding of the world. In some shape or form, denial tends to some degree to be a permanent feature of human life. Thus it is axiomatic in psychotherapy that 'we all need our defences'. As T.S. Eliot says 'Humankind cannot bear very much reality'.

But just what are the realities that we cannot bear?

The collective attitude towards climate change is constantly shifting. Thus this description written in January 2007 may be, when published, quite inaccurate and out of date. In September 2005 the central example of denial was the refusal to believe that climate change was happening at all. That has changed. Thanks largely to a remarkable transformation in the mass media the reality of climate change now seems to be almost universally acknowledged. But this acknowledgement is little more than lip service. Denial in fact continues, and takes the form of an extraordinary muddle of confused and contradictory assertions. What is now denied is that it is due to human action; or that there is any hurry to avert it; or that there is still time to avert it; or that the mass of ordinary people will ever be willing to accept the necessary changes. Above all what is denied is the reality of our feelings about it.

But underlying all this is a deeper reality: namely that *our saving the climate is impossible without a radical reduction in our standard of living in the rich North*. It is this that we cannot bear, and that no government, no political party, dare acknowledge. It is indeed hard to bear or understand for those under 50, who have never known anything but increasing affluence.

This state of collective confusion stems from an almost complete absence of the clear leadership essential to a mass movement. Most people assume that it is for Government to tell us what needs to be done, and how we should do it. But Government dare not undertake this task in any detail, for fear of the electoral consequences. The resultant vacuum is filled by this hubbub of contradictory voices. The failure in leadership is repeated right across our culture. Since the task before us is supremely spiritual we might expect leadership from the churches. But that too is almost totally lacking.

Our denial of the full horror of the situation and our feelings about it tend at times, with the best intentions, to be deliberately reinforced. I was recently at a meeting addressed by a very senior member of the IPCC. In a brief moment of private conversation, I asked him if he was optimistic about the aversion of climate catastrophe. 'No', he said, 'I'm pessimistic.' Yet this was not the mood or the conclusion which he had conveyed in public half an hour before. And those who shared the platform with him had been at pains to emphasise how important it was, in presenting the situation to the public, not to be 'negative', not to indulge in what they called 'doom and gloom.' It is widely assumed, with some reason, that as a culture we have lost the ability to confront harsh reality. Here again the churches have much to answer for.

How is Gaian theory related to all this? In her introduction Mary Midgley refers to the hard work of addressing 'the whole relation between our inner and outer lives'. In our meetings of the Gaia Network we have discovered just how hard that work can be. We have had great difficulty in expressing and sharing our feelings. Our discussions have remained almost wholly at an intellectual level. Climate change is imminent, and is the greatest threat mankind has ever faced. But in our discussions its impact upon our inner lives and feelings has been almost totally ignored.

But the feelings are certainly *there*, beneath the surface. From time to time, in isolated privacy, far away from the meetings, they break through the denial, and suddenly erupt with great passion and intensity—often in the form of grief and utter despair about the prospect before us and our children. But not being shared they remain unrelated to the intellectual context in which they need discussion. So the energy in them, which might be transformed and mobilised to promote the collective action of a mass movement, instead goes to waste.

All this suggests that, contrary to expectation, an academic interest in Gaia provides in itself no immunity from denial. On the contrary, we seem to have here confirmation of the words of the World Council of Churches:

> The threat of climate change is of such magnitude that it surpasses the human capacity to react. People tend therefore to protect themselves by pursuing their present way of life.

For an academic, writing a paper about Gaia while avoiding discussion of one's feelings is indeed 'pursuing one's present way of life'! It leaves one in denial, well protected from the inner and underlying realities. If the message of Gaia has validity and potency, it reaches us not through the academic intellect but rather through the vulnerable human heart. Ideally, it reaches us through the communion of heart and head together.

Last, but very far from least, denial ensures that we are not much troubled by ethical considerations. According to Jung, we have access to these through our feelings. What we in the rich countries are doing to the planet, to people in poor countries, and to future generations, is, beyond all question, morally outrageous. It amounts to a holocaust. But thanks to denial, all that, together with our feelings, is virtually ignored.

Let us return to the words of Joanna Macey:

> The very danger signals that should rivet our attention, summon up the blood and *bond us in collective action* tend to have the opposite effect. They make us want to pull down the blinds and busy ourselves with other things.

That is the tendency that at present undermines all our efforts. It is that that we must consciously combat and overcome. We must, that is to say, deliberately set out to 'bond together' in collective action, to meet together, in small groups, and discuss both our understanding of the situation, and our feelings about it, and how we are to achieve our aims. That means sharing those feelings—of despair, impotence, anger, grief, depression, etc. In this matter it is fatal, spiritually, psychologically, and politically, to be isolated. Dr. Rowan Williams described this well when he spoke, in his Environment Lecture, of the need to overcome denial by promoting 'new and secure relationships enabling us to confront unwelcome truths without the fear of being destroyed by them'. It is our fear of facing, in isolation, the 'unwelcome truths' that drives us into denial.

We are certainly perilously near a 'tipping point' in terms of climate change. It seems possible that we are also near a 'tipping point' in terms of public awareness and readiness to act. What is now needed — and with great urgency — is the establishment of that capacity for collective action of which Joanna Macey speaks, and which at present is rendered impotent by the psychologically divisive effects of denial. We need to use that collective action to put massive pressure on politicians to take those measures which can yet save our civilization and the future of human life.

David Midgley

Climate Change and Spiritual Transformation

The issue of climate change has given rise to an altogether unprecedented level of public concern and activity on the part of governmental and non-governmental organisations of various kinds; yet there appears no sign as yet of the emergence of the political will necessary to convert this concern into action on anything like the required scale. How is it that we appear to be unable to change course, despite mounting evidence that we are headed for a disaster of unparalleled dimensions? The root of the problem seems to lie in a pathological condition of the human psyche; and this psycho-pathology needs to be understood at a collective rather than an individual level. It is the mind of our civilisation which is sick, and the structures of neurosis which Freud discovered within the psyche of the typical member of that civilization, resulting in systematically irrational and destructive patterns of behaviour, are a reflection, in microcosm, of structural and systemic distortions and imbalances in the psyche of the society as a whole.

Among the babel of voices in the consciousness of an individual, the voice of wisdom can be clearly discerned, but often does not prevail over irrational and unconscious motives which the individual does not fully understand — despite an uneasy awareness that the price of ignoring that quiet voice of warning may, in the final event, be one which he or she will not want to pay. It is the same with the mind of a society (what we call culture) — the warnings are heard, the attempts to dismiss them are unconvincing, but the courageous act of fully acknowledging the danger and making the necessary change of course seems to be too difficult and uncomfortable to contemplate. We are a society in denial, addicted to consumption and growth, locked into a mechanistic paradigm which persistently attempts to cure the disease by prescribing more of the cause.

I think that the cultural disorders which underlie the present ecological crisis need to be understood on three levels. In the first place, we need to address a collective psychopathology very strongly analogous to addiction to a drug such as heroin — and here, of course, as with an individual addict, rational persuasion is unlikely to be successful on its own. Secondly, and inseparably bound up with this pathology, the world-view which has guided the development of our industrial civilization for the last four centuries has now ceased to serve us; we are at a point of paradigm change comparable to the transition from Aristotelian to Newtonian physics, and if anything more far-reaching in its implications. Lastly, as with that earlier transition, the present re-orientation away from a mechanistic world-view and a fossil-fuelled industrial mode of production will necessarily be accompanied, and partly motivated, by a profound spiritual re-evaluation. This is in fact the deepest layer of the psychic shift we are witnessing, and the one which most radically challenges the existing modes of consciousness that are the source of the problem. I would like to look at each of these in turn, in order to advance progressively to the heart of the issue.

First, then, our practical orientation to the world, as a society, is dysfunctional, distorted by our massive dependence on ever-increasing quantities of fossil fuel energy. The balance between effort and reward that existed in pre-industrial societies (though not for the ruling élite) fulfilled a significant ecological and psychological function. This balance has been destroyed by the discovery of the means to use fossil fuel energy to extract more fossil fuel energy, and harness it to replace human labour. The hyper-exponential growth of this process has persisted now for long enough for it to be perceived, in a bizarre paradox, as a stable condition — to an ecologist an obvious contradiction, but apparently not so to an economist.

As we entered into a closer and closer symbiosis with industrial technology, our symbiotic relationships with the wider ecosystem were ruptured, damaging our physical and mental health as well as the health of the ecosystem. Coal- and oil-fuelled technology provided a short-cut route to physical comfort and satisfaction, in much the same way as heroin and similar drugs provide the addict with a short-cut route to a temporary experience of bliss by overdrawing on the body's physiological reserves. In both cases, the result is progressively increasing damage to the complex system of equilibria which maintain the integrity of the organism/ecosystem. In seeking a better understanding of this process we need to make use of the tools of cybernetics and systems theory, but also those of psychoanalysis and other psychological disciplines. Addiction is both a physiological and a psychological condition, and in attempting to find a remedy for its socio-economic analogue, we need to look at how we can repair dysfunctional motivational patterns as well as changing the metabolic patterns of our economy.

Inextricably linked with these destructive habits of behaviour and motivation is an anachronistic set of habits of thought: the paradigm, or world-view, of mechanistic, reductionist science. This conceptual scheme was extremely useful in the transition from an agrarian subsistence economy to an affluent industrial one, but has become dangerously inappropriate today, when our impact on the natural systems on which our life depends has become inconceivably greater than it was when the paradigm arose in the seventeenth century. The mechanical model of reality developed by Galileo, Descartes, Newton and others has been spectacularly successful in explaining and manipulating a certain limited range of phenomena, but the attempt to extend it to a comprehensive account of the nature of living processes has resulted only in caricature. Worse, the use of a paradigm that is deeply grounded in a conception of matter as essentially inanimate and dead, to guide us in our interactions with the living world we actually have to deal with, has inevitably resulted in our inflicting profound and increasing damage upon the fabric of that world, on the body of Gaia our mother, of whose very existence as a living entity we have been fatally unaware.

While the world of 'Big Science' devotes itself to the twin misconceived, hubristic reductive projects of seeking the ultimate explanation of Life in terms of random molecular mutation combined with natural selection, and the ultimate explanation of the entire physical universe in a supposed 'Grand Theory of Everything', the real creative thrust of science continues on the sidelines, in the advancing understanding of complex systems through systems theory, complexity theory, Gaia Theory and the like. All these new disciplines share a conception of causality based on mutual interaction between all the parts of a system, and between part and whole, in place of the mechanistic model of linear causality.

The essential relationship between the intellectual and motivational aspects of our collective mind-set can be summed up in the observation that a materialistic conception of the nature of reality induces, and is induced by, a materialistic value-system. Our intellectual perspective and our set of values and priorities are both manifestations of our fundamental existential orientation to the world, and to ourselves and our fellow beings. Our notion of who and what we are, of the essential nature of our relationship to others, and of the cosmic whole in which we are embedded, form the underlying core of our moral, intellectual and emotional life: this is the domain of the spiritual, and it is here that we find the root of the manifold physical, emotional and intellectual disturbances with which our civilization is grappling.

The scientific revolution of the seventeenth century grew out of a reaction against an excessive form of idealism — the mediaeval Platonistic interpretation of Christianity which besought believers to disregard the sufferings of this life in the expectation of an eternal reward in a transcendent, ideal realm in the next. The negation of the reality of material suffering, and of the autonomy of the individual, came into conflict with the needs of the time, when improved modes of production and communication made possible a rapid, and much-needed, advancement in the material conditions of life and in the general level of education. Materialism, individualism, and rationalism became the motive force of European civilization through the Enlightenment and the Industrial Revolution, accompanied by a gradual and progressive erosion of traditional Christian spirituality. In a sense the root cause of the problem lay within that spiritual framework itself: it was the dualistic split between the realm of Spirit and the realm of Matter posited by this world-view which presented European thought with a dichotomous choice between assigning reality and value to one of these two realms, and denying it to the other. Descartes' attempt to resolve the problem by assigning equal reality to both, while maintaining their complete independence and irreconcilably different natures, places him at the pivot-point of this movement of thought; but his solution was incoherent and unstable, and subsequent history confirmed the trend to progressively more extreme denial of the reality of the realm of Spirit.

As the intellectual and economic project of scientific materialism and industrial production reached its zenith in the twentieth century, the disastrous results of such neglect of the spiritual side of life became apparent in the phenomena of total war and totalitarian political systems. Imperfect as mediaeval Christianity was as a spiritual foundation for the moral order, in the austere philosophy of atheistic materialism such a foundation was almost entirely lacking, and the intellectual landscape of twentieth century moral philosophy was dominated by various forms of ethical relativism and nihilism. Both philosophies, moreover, shared a deficiency in the area of reverence for Nature, and the Baconian project of 'conquering and subduing Nature' echoes Jehovah's injunction to Adam to 'Subdue the earth, and have dominion ... over every living thing that moveth upon the earth.'

Though its roots are discernible earlier, the inevitable countermovement to this extreme condition of culture first became visible in the 1950s, with the emergence of the Beat generation in the United States. The attempt to forge a cultural and spiritual alternative to both dogmatic patriarchal Christianity and rationalist scientific materialism was linked with the onset of the transmission of Buddhism to the West, consequent on America's wartime involvement with Japan and its post-war economic reconstruction. Buddhism provided a Middle Way between the extremes of absolutist idealism and nihilistic materialism, an alternative rationality that transcended rigid logic but respected the autonomy of the individual and retained a sensitivity to the values of Life and Nature. For a significant cultural minority who continued to seek out a path different from the *status quo ante*, it gradually displaced the radical iconoclasm of the Beats and Hippies as the focus of a new spiritual ethos.

The classical tradition of Buddhism shares with the emerging disciplines of holistic science referred to earlier a philosophical view of reality centred on the idea of interdependence. Its profound analysis of the relationship between consciousness and reality, and deeply-grounded tradition of spiritual practice, complemented the new understanding of Nature provided by systems science and ecology to offer an overall vision of life with the potential to form the foundation of a new movement of global culture. A third central component of this emerging constellation of ideas and praxis was the towering spiritual genius of Gandhi, who showed by his inspirational example that uncompromising moral idealism and selflessness were entirely compatible with political effectiveness. His unique brand of engaged spirituality seemed to unite seamlessly the Western ideals of radical social and political reform, derived from Tolstoy's Christian socialism and Kropotkin's humanistic anarchism, with the timeless values of non-violence, reverence for Life and devotion to God that he found in classical Hinduism.

In this convergence of spiritual, philosophical and scientific ideas, which for convenience we can label Deep Ecology, I believe we have the necessary intellectual and therapeutic tools with which to attempt the task of cultural transformation needed to address the present ecological crisis, of which Global Warming has become the universally recognised symbol. The essentials of this philosophy first came clearly into focus in 1973 — the year of the first Middle East Oil Crisis — with the publication of Donella H. Meadows' report to the Club of Rome, *The Limits to Growth*, and of E.F. Schumacher's classic *Small is Beautiful*. Meadows' report was the first (and remains the most significant) attempt to articulate in quantitative detail the dynamics of the total system representing the interactions between the human economy and its natural environment, while Schumacher's brilliant critique of the underlying pathology of culture responsible for the crisis, the twin diseases of industrial gigantism and mass consumerism, built on the spiritual foundations laid by Gandhi to set out a path towards a more modest, balanced and spiritually sane social and economic order.

The warnings were largely ignored, however, and today we are confronted with the spectacle of enormously powerful global institutions, invested with the power of decision and the responsibility for setting the policies which will determine our future, resolutely ignoring or actively denying the increasingly self-evident destructive consequences of our present path. To someone whose basic orientation to the world is grounded in the ecological paradigm, this is apt to induce a mixture of rage, despair and mental paralysis. It may be, however, that it is in this very pattern of increasing rigidity, this closing of ranks and escalating denial, that we can discern the seeds of hope. The process of rigidification and denial is typical of the terminal phase in the life-cycle of a paradigm, and might be compared to the chrysalis stage in the life-cycle of a moth or butterfly: it could precede a very rapid and radical transformation in the intellectual, social and economic structures of the society. While the outer shell of the organism seems rigid and immoveable, invisible changes are taking place within which may erupt dramatically when they reach a critical stage of development.

Such a paradoxical-seeming process of transformation might be difficult to imagine within the conceptual framework of the earlier paradigm, but is perfectly in keeping with the understanding of how complex systems evolve which the new paradigm offers. In a systems-theoretic sense, it appears analogous to the process of spontaneous remission that is observed in cases of 'terminal' cancer, and also with recovery from severe heroin addiction. Conventional, interventionist medical or psychotherapeutic techniques are often powerless in such cases, where recovery comes about through a holistic response of the organism / psyche to a crisis of survival. In this analogy we can compare the situation of those of us who are committed to living and thinking in alignment with an ecological, holistic understanding of the world, with that of the healthy cells in a body threatened by cancer. The breakdown of the 'old order', the structural systemic condition which precipitated the cancer, creates an opportunity for a restructuring in which the life-affirming tendencies represented by the healthy cells are able to re-assert themselves and become the dominant influence shaping the life processes within the organism. On the spiritual, psychic level this is paralleled by the experience of the 'Dark Night of the Soul' described by St. John of the Cross, in which, when the resources of the conscious, rational mind reach their limit in the attempt to purify the self of its tenacious attachment to destructive modes of being, an encounter with a death-like condition of spiritual desolation turns out to be the prelude to a new dawn of awakening and healing.

I think that this representation of our present predicament has the important merit of striking the correct balance between optimism and pessimism. A crisis of this nature poses an absolutely real danger—it would be fatal to adopt a shallow interpretation that views the transformation I have pointed to as a deterministic process with an inevitable positive outcome. At the same time, the nature of the systemic processes involved means that, despite the depth and magnitude of the problem, there are real and tangible grounds for hope, not merely for a degree of alleviation of our condition but for fundamental positive changes that would, in a profound sense, make all of the suffering worthwhile. The outcome will depend on our conscious actions and choices, and the present generation bears a momentous responsibility for navigating the human species through this perilous stage of our collective journey. But we must somehow learn to bear this responsibility lightly, for if we allow ourselves to be emotionally crushed by its weight, we cannot fulfil the task which our place in history has assigned to us. In achieving this balance I believe the crucial spiritual attribute that we need is *vision*—if we can hold consistently before us a clear and coherent vision of the condition of the world that we seek to bring to birth, to hold on to this vision in the face of the challenges of cynicism and despair induced by the darkness about us, and persist in our efforts to communicate this vision to those who have not yet awakened to the new possibilities, we may before it is too late succeed in igniting a process of social, intellectual and cultural change that will transform the human scene more rapidly than we can readily imagine or believe.

The spiritual importance and power of vision is recognised in the teachings of Buddhist Tantra, in the principle known as *bringing the future result into the present path.* By constantly redirecting our attention to a clear vision of the enlightened state that we wish to actualise, it is understood that we are actually bringing that result into being. According to the Buddhist theory of interdependent origination referred to earlier, the unfolding of the historical processes which create our future are not determined by some lifeless principle of random chance or mechanical causation, but are the result of a creative process in which consciousness depicts possible future states of affairs and orients itself towards those imagined worlds. As long as we seek to understand consciousness and its relation to reality in a mechanistic fashion, we will not be able to comprehend the power that a change in our world-view has to bring about a change in our world.

Susan Canney

Reconnecting a Divided World

Links Between the Global and the Local

'If the elephants disappear we know that
the land is no longer good for humans.'
Malian villager

This statement struck me with force. The reason is that it is so counter to the attitude that I am used to hearing in my work of
nature-conservation-planning, where nature tends to be regarded as
a cost, an expensive luxury, for which we have to make sacrifices.
One can quibble with the details of the argument but the emergent
truth remains, starkly and simply put: the disappearance of elephants would signify the effect of the very processes that are making
it more and more difficult for humans to live in this area.

For me it sums up a mind-set that regards the land as primary, that
sees the health of individuals and society as impossible without the
health of the natural support systems, as it acknowledges their deep
inter-dependence. It stands in contrast to that of our industrial economic society that has lost a sense of the 'web of life' and seeks to
pick and choose which bits of it are most useful and which bits can be
disposed of.

The Outdated View and Why It's a Problem

We are increasingly aware that there are several crises looming —
energy, climate change, damage to ecosystems, population — and we
don't know how to deal with them. Most of these problems spring
from the crisis of perception outlined by Mary Midgley in the introduction to this book: the misperception of our relationship to the
planet, and to each other, that filters through to how we construct

our lives. As she explains in *Science and Poetry*, 'our visions, our ways of imagining the world are expressed in all our thoughts and actions.'

She refers to the set of imaginative habits that has been associated with the practice of science since the mid-sixteenth century. This mechanist world-view, with its atomistic reductionist approach, is behind the compartmentalization of thought into the academic disciplines, and has led us to deal with problems in isolation from their wider context. While this has enabled precision in thinking and the development of ingenious technologies, it has meant we have lost sight of the 'big picture'. It has enabled us to displace the impacts of our activities out of our sphere of consciousness and onto the environment, other parts of the world, and the less powerful, until these accumulate in scale or in combination to manifest as a larger, unexpected problem. We then deal with the new problem in the same way, until we come to the point we are at now where we have reached the scale of the planet and its physical limits. We're now affecting the global cycles of carbon, nitrogen, phosphorus and sulphur, and as Lovelock describes, 'by adding greenhouse gases to the air and by replacing natural ecosystems with farmland we are hitting the Earth with a 'double whammy'. We are interfering with temperature regulation by turning up the heat and then simultaneously removing the natural systems that help regulate it'.

This old view of the world leads us to respond to unfolding threats using what Homer-Dixon refers to as a 'management approach' that consists first of denial, followed by reluctant management once the evidence can no longer be ignored. It assumes that there is nothing that is too hard for our brains to solve or that science, markets and democracy cannot deliver when needed. It typically involves isolating the problem, analyzing the data, forecasting the future by extrapolating from the past, and laying out detailed policies to reduce the problem's seriousness and adapt to its consequences.

But we are now realizing that the world does not work in the way assumed by this approach. As Goodwin describes in this book, a new vision of the world has been emerging in the sciences through the twentieth century, and yet our perspectives, as well as our organizations, institutions and procedures are modeled on the old.

The Gaian World-view

The new vision senses the Earth as a complex system, Gaia, and recognizes that our globalized social world is reliant on the natural world: when there's trouble in nature, there's trouble in society. This means that Gaia, the planetary system, has emergent properties that cannot be attributed to any particular part but only to the system as a whole. It means that incremental small changes can accumulate, or diverse impacts converge, to suddenly shift the behaviour of the whole system to a radically new mode. It means that the behaviour of such a system is contingent on a host of factors, large and small, knowable and unknowable, that are beyond our brains to predict or control precisely. Constant change and surprise are therefore inevitable.

Accepting this requires a corrective dose of ecological humility (see Anne Primavesi and Stephan Harding this volume). But even though we can't make exact predictions about detailed events, we are not impotent. We can still forecast general trends, and try to anticipate harmful outcomes in the future by better understanding the pressures affecting the world and how they might act, singly or together. But to do this requires a different approach.

We can also recognize that some of our systems are not adequate to cope with the new vision of the world, nor with the problems that are emerging as a result. Homer-Dixon continues his discussion of the management approach by pointing out that addressing the underlying causes of our hardest problems usually runs headlong into staunch opposition from the status quo and its powerful interest groups, but that small breakdowns can be opportunities for renewal and betterment. A rigid and highly compartmentalized approach, however, makes it more difficult to harness these opportunities because we fail to recognize the warning signs, understand the underlying causes, or imagine and implement more radical and far-reaching solutions.

More fundamentally, we need to take a leaf out of nature's book and constrain the shock of sudden change by making our societies and our systems — food-supply, energy, transportation, financial — and each one of us more resilient and more supple in response to rapid change, even if it costs more and we have to give up extra efficiency and productivity as defined by narrow economic metrics.

Can Gaia Help?

Gaia helps us to make sense of the world at a planetary level but can it help at the scale of our everyday lives and in our day-to-day actions and decisions? Can it guide us despite the fact that our work-places and institutions function according to the concepts of an outdated-world-view? I believe it can. We can begin by restoring the connections to see the whole picture. In the personal account that follows I describe an example of how a Gaian approach transformed my insight and understanding of a particular situation.

I call this a Gaian approach because it sought a unity of knowledge through a top-down perspective, while retaining a groundedness in physical reality. By reinstating the wider context it aimed to map out the system-architecture to integrate the precision of the disciplines and their multiple perspectives into a system-wide vision.

Some years ago I found myself walking into a highly polarized and apparently irreconcilable debate about the impact of pastoralists in an East African Game Reserve and whether they should be allowed there. On one side were the predominantly natural scientists and Reserve managers whose focus was the need to protect wildlife from the resource demands of rising human populations and their associated environmental damage. On the other were the predominantly social scientists who maintained that perceptions of environmental damage were unnecessarily exaggerated to justify land alienation, and that pastoralist methods of herding had evolved to be in harmony with nature. Each side was genuine in intention and had amassed evidence to support their case.

The two sides drew largely on two separate bodies of literature that hardly engaged with each other. The reductive methods of scientific inquiry meant that many of the accounts were partial and gleaned by carving out a part of the system, and divorcing it from its context to devise a snapshot with reassuring quantities attached. By selecting appropriate snapshots, one could construct a persuasive argument, and the motivation to do so was amplified by the politically charged nature of the situation. Trying to piece together a sense of the whole was much more difficult. It involved looking for unity rather than division, which meant expanding the context to embrace the differences in knowledge, understanding, preconceptions, and priorities of the authors. The process was iterative, feeling my way around the system as I combined the information of my senses with the observations of others until the fragments of knowledge began to link up and a unified vision emerged, a bit like the sensation of suddenly feeling a three-dimensional figure lift out of the page of apparently random dots.

What emerged was a sense of the savanna environment as an ever-shifting mosaic of woody plants and grass responding to and affecting the interaction of landscape, moisture, nutrients, fire and herbivory. Its diversity and resilience arose from its variability.

Rainfall, for example, varied from year to year and between alternating wet and dry seasons generated by the passing of the intertropical convergence zone, a result of the tilt of the earth in relation to the sun.

Thus I charted how rainfall varied over the land surface, how it responded to the hills and the plains, how the characteristics of the soil emerged and all together influenced the water available for plants. I found system thresholds: where the annual rainfall was less than around 600–800mm life was water-limited, while at higher values, it was nutrients that became the limiting factor. I learned how the soil properties influenced what grew, as for example where rapidly draining, nutrient-poor, sandy soils tended to support trees whose roots could access the water table, while clay soils in the drainage-ways could only grow grass. How fire encouraged grass and discouraged trees yet was itself influenced by rainfall as good wet seasons produced a high fuel load and hot, powerful fires. Extensive grass plains supported grazers like gazelles and zebra, who in turn encouraged grass growth, unless present in large numbers when they reduced the grass available to burn, and the woody vegetation could gain some ground at the grass's expense. Several fire-free years could allow the woody canopy to thicken and, starved of light, the grass thinned and fire could no longer penetrate, unless particularly hot or until elephants opened up the thicket in their search for forage.

The value of this particular piece of savanna was as a wet-season dispersal area for the wider Tsavo ecosystem. It was relatively well-watered in the rains due to adjacent mountains but waterless in the dry season. Wildlife migrated there to take advantage of its rich forage when surface water was abundant, giving areas close to permanent water — and heavily used in the dry season — a chance to recover. The lack of surface water meant that the Reserve had been little used by humans, but since its gazettement in 1951, the Department of Wildlife had dug three dams to encourage wildlife viewing and tourism revenue. This made the Reserve attractive to pastoralists, who had been displaced from traditional lands elsewhere by agricultural land-take. Their numbers steadily increased during the 1970s and 1980s until a particularly determined Regional Wildlife Officer managed to overcome local political alliances and evict around 13–18,000 people and their cattle in 1987–8.

The concern over livestock was that large numbers of livestock displaced wildlife; and that heavy, sustained grazing by one herbivore species was an atypical disturbance that the system had not evolved to cope with. It could therefore reduce the diversity and resilience of the ecosystem, pushing it into a state of low productivity from which recovery may not be possible on human timescales.

By comparing satellite images taken at key points in the Reserve's history — before the influx of pastoralists, at eviction and present day — it was possible to identify the vegetation changes that had occurred at each point in the Reserve, and compare them with the equivalent changes in the density of human occupation and livestock as obtained from aerial photography and census statistics (while checking that other factors, such as the pattern of rainfall and elephant populations would not have been able to produce these changes).

There was a correlation. The impact of the pastoralists varied across the area according to the intensity of their use modified in various ways by the vulnerability of the habitat. In some areas human impact could be regarded as within natural variability (although it is unknown whether this would have continued to be the case if the pressure had not been removed). In these areas it could be regarded as a matter of preference whether one lamented the loss of woody vegetation because it rendered the vistas more uniform, or welcomed the increased visibility and larger numbers of animals. Other impacts clearly reduced the resilience of the ecosystem such as over-frequent burning that ate into the forests of the hills, rendering the slopes vulnerable to loss of top-soil during the first rains; or where sustained, intense use had caused an irreversible loss of vegetation.

At the same time it was also clear that the pastoralists were not the ultimate cause of the problem. Traditional peoples know much more about living within natural systems than we do. Directly dependent on their environment, traditional systems generally aim to preserve its health. (By 'health' I mean a system that exhibits such qualities as high diversity, productivity and resilience — see Goodwin for further discussion).The traditional migrations and livestock management practices of the pastoralists had evolved to cope with the limits of the savanna environment and functioned as part of the ecosystem, but over the past decades, expansion of agriculture and other human activities in the east African savannas has led to a decline in the extent of both pastoral and wildlife areas. This has resulted in the restriction of wildlife to protected areas, while herders and their livestock have to be sustained on smaller areas of pastoral land, increasingly forced to settle, turning to agriculture and wage labour (often as herdsmen for the large herds of rich, urbanised pastoralists) as they are drawn into the wider economy. Pastoralist impact was a product of the wider, societal forces driving land alienation. Both they and the environment were victims of the modern world.

The aim of this chapter was to make links between the global and the local, and to flag how the trends and processes that we observe at both scales arise from the same divisive thought habits that we carry into all scales of activity. That our resilience depends on Gaia's resilience was the understanding behind the statement that opened this chapter. Resilience is an emergent property that results from a synergy between the parts of a system. It therefore needs to be built in at all scales. The implication for our current situation and how we act within it is that there are no magic bullets or technologies that will fix all our problems. We need to be active on multiple fronts, guided by a true appreciation of our place on Earth.

Sources

Behnke, R.H., I. Scoones, C. Kerven, Eds. (1993), *Range Ecology at Disequilibrium: New Models of Natural Variability and Pastoral Adaptation in African Savannas* (London: Overseas Development Institute).

Brockington, D. (1998), *Land loss and livelihoods: effects of eviction on pastoralists moved from Mkomazi Game Reserve, Tanzania*. PhD thesis, University of London.

Canney, S.M. (2001), *Satellite mapping of vegetation change: human impact in an East African semi-arid savanna* (Lymington, UK: Pisces Conservation Ltd.).

Capra, F. (1996), *The Web of Life* (Harper Collins).

Coe, M.J., N.C. McWilliam, G.N. Stone and M.J. Packer, Eds. (1999), *Mkomazi: the Ecology, Biodiversity and Conservation of a Tanzanian Savanna* (London: Royal Geographical Society with The Institute of British Geographers).

Holling, C.S. (2001), 'Understanding the complexity of economic, ecological and social systems', *Ecosystems*, **4**, pp. 390–405.

Homer-Dixon, T.F. (2006), *The Upside of Down: Catastrophe, Creativity and the Renewal of Civilization* (Island Press).

Illius, A. W. and T. G. O'Connor (1999), 'On the relevance of nonequilibrium concepts to arid and semiarid grazing systems', *Ecological Applications*, **9** (3), pp. 798–813.

Lovelock, J.E. (2006), *The Revenge of Gaia* (Penguin Books).

Lovelock, J.E. (2005), *Gaia: Medicine for an Ailing Planet* (Gaia Books).

Midgley, M. (2001), *Gaia: the next big idea* (Demos).

Midgley, M. (2001), *Science and Poetry* (Routledge).

Maggie Gee

Imagining Gaia

Art Living Lightly With Science

It's well-known that the name of the Gaia theory came from conversations between the scientist James Lovelock and the novelist William Golding in 1967, when they were neighbours in Bowerchalke, Wiltshire. They would walk the half-mile to the village post-office, but when the conversation was exciting they would go far beyond. Fortunately, Golding knew enough about physics — culturally ambidextrous, he had changed from Natural Sciences to English at Oxford — to be fascinated by Lovelock's theory. But he also understood the power of language and told his friend he had 'better give it a proper name'. Why not Gaia?

Earth as a single mythic figure, the Greek goddess, first to be born after Chaos, mother of all that lives: the metaphor came naturally to an artist like Golding, and it proved a powerful way of introducing Lovelock's complex idea of a self-regulating planet into the popular imagination. Even though, as Mary Midgley points out in her introduction to this book, James Lovelock's ideas had to be freeze-dried as 'earth science' for academic respectability, 'Gaia' worked for the wider audience. Science benefited from the seductiveness of art, its offer of play and pleasure, its appeal to different parts of the brain which enjoy narrative and visualisation. Of course this symbiosis has also enlivened the popular science books of the last three or four decades — Lovelock generously praised what he called the 'beautiful metaphor' behind Richard Dawkins's *The Selfish Gene*.

It's a two-way traffic. As a novelist, I need ideas as well as images. Science feeds my imagination. When I was a child, I lived in the country, and wrote poems about primroses in spring, corn and wheat in summer, autumn leaves. But the new seasons are confused. A yellow rose is in bloom outside my window on this January day, side by side with a winter hellebore. As an urban adult, I need to understand. Science as well as poetry.

For reasons to do with class and money that were commonplace at the time, neither of my clever parents had quite the education they deserved, but within the limits of that, it was a marriage between a scientist and an artist. My mother adored jokes, puns, rhymes and stories. My father, by contrast, was a wizard at maths, liked order in everything and had the remorseless logic of a scientist. The divide continued in their children: my elder brother read physics, I read English.

At Oxford, through my brother, I got to know, and was lucky enough to be befriended by, a remarkable couple of intellectuals, the ecologist C.S. Elton and the poet E.J. Scovell. Though all marriages are mysterious and none is perfect, they were for me an exemplary pairing of art and science. Charles Elton has been seen as the founding father of British ecology, and Robert May said in 1997 that 'to my mind Elton is far and away, clearly, the leading figure in ecology in the world in the first half of the century'. Books like *Animal Ecology* (1927, reprinted nine times and re-published in 2001) and *The Pattern of Animal Communities* (1966) were milestones in the understanding of the interrelationship of species. He told us that nothing lives alone: the fluctuation of one animal population was affected by climatic factors as well as by the fluctuation of other populations, chiefly its enemies — 'the arctic fox fluctuates with the lemmings, the red fox with mice and rabbits ... and so on.' Charles studied with patient concentration the interconnected lives of domestic species in nearby Wytham Woods, but he also travelled worldwide, to icy Spitsbergen and tropical Panama, recording the subtle rhythms of different eco-systems, pioneering in particular the use of statistics (or, to give it a more friendly name, counting) in understanding what was happening in nature. He was a deeply shy man, not clubbable, though as Director of the Bureau of Animal Ecology he was known as 'The Boss', and he was the first head of the Nature Conservancy Council.

His wife, E.J. Scovell, was equally remarkable as an artist. Joy Scovell sometimes worked as Charles's field assistant, but she was also a lyric poet praised by the acerbic Geoffrey Grigson as 'the purest of women poets of our time'. Her clear, exact and economical nature poems express her love for the natural world in a different, but cognate, way to Charles's work, and she often uses nature as a metaphor. In 'The Clover Fields' she is writing about a particular landscape but also about the illumination that sometimes compensates a thinker late in life:

> The fields are overcast with light at evening,
> With marguerites increased, a chalk-white settling.

I was very young, around 20, when I first went to lunch with Charles Elton and Joy Scovell in their tall narrow house in the middle of Park Town's furthest crescent, its whole length alive with Virginia creeper. Charles enjoyed gardening enough to look after the communal garden cupped by the crescent, 'The Jungle', as well as his own, and believed in being hands-on, saying briskly that 'the outlook of the naturalist [must] leaven the crust of science'. Lunch was always pleasantly austere, soup followed by Bath Olivers and cheddar and apples, taken in the basement kitchen into which the sun streamed, throwing shadows from the barred windows and the blowing creeper leaves on to the oilcloth, lighting up a vase of silver Honesty on the window-sill. Though already nearly seventy, Charles seemed curiously ageless, fast and light-footed on the stairs despite his weight. He had large pale eyes behind elliptical glasses and a very quiet voice that compelled attention. Joy was slender and kept her grey hair long and loosely caught up at her neck because Charles liked it that way; her blue gaze was often thoughtfully averted. They were anti-consumerists *avant la lettre*, driving (when they absolutely had to) a thirteen-year-old Morris Traveller trimmed with weathered wood.

What did I learn from them? That science and art were two sides of the same observant curiosity, both activities impelled by a desire to understand, both finding and recording patterns in what at first seems random and chaotic, both finding the universal in the particular. Charles knew this was not easy:

> one has to make a considerable effort to break through from one level of study to another above it, ... to forge through the apparent complexities to a higher level of integration and arrive at simple ideas applicable to the higher level but invisible from the jungle below.

He might equally well have been describing the struggle of a poet or novelist to find and make the over-arching metaphor of the work from its individual insights. I learned that scientists and artists need each other — these two eager minds moved in five seconds from talk about the migrations of sea-birds to the poet Peter Levi's book, *The Shearwaters*.

I learned caution about how much can be changed by statements and petitions. Already in the late 1960s and early 1970s Charles took a sombrely apocalyptic view of the damage that excessive human numbers would do to the planetary ecosystem. Having pioneered the idea of conservation, he became, in private, a pessimist about environmental campaigns. Charles's gift was to bring scientific rigour to the field of natural history and to spread a radical new understanding of the planet's delicately balanced living systems, not to be a cheer-leader.

We all work within the limits of the psyche we were born with. I have always been over-susceptible to fear (which does not stop me being rash), but fear can also be a gift, acting as an imaginative gateway from the present to the future. Because we are short-lived creatures, the chances of any one of us experiencing major disasters has until recently been relatively small. However, in the longer term we know the human species has come through all manner of catastrophic events, from meteorite impacts to tidal waves and ice ages. My fear-proneness means that the shadow cast by this knowledge on the sunny everyday is more distinct than it probably is for innately calmer people. And it drives me to tell apocalyptic stories; but also to imagine redemptive endings, ways of surviving. My fascination with science helps to contain both hope and fear within a continuum of rationality.

The author of *Voles, Mice and Lemmings* (1942) taught me that populations surge, and then crash. Disease or a plague of enemies will tend to carry off an impossibly inflated population when the conditions are right, or else the animals or plants which are their food source will be exhausted. Perhaps bird flu or famine will do better than treaties or protocols at making twenty-first-century human beings cut down their carbon and methane emissions. After all, according to William Ruddiman's new book, *Plows, Plagues and Petroleum,* steep declines in the CO_2 record tend, historically, to follow major plague pandemics, when farmland reverts to forest. (But of course, no one wants climate control by lethal disease. Can we really only learn through hideous disasters that tear apart the small, ordinary family affections and patterns of work that make human life worthwhile?)

When I was writing my third novel, *Light Years,* in 1984, I had recently married, and we were living in a couple of rooms near Regent's Park. Lovelock's *Gaia: A New Look at Life on Earth* had been published in 1979. My husband had bought a second-hand copy in the early '80s, and we were both intuitively taken by the central idea of the earth as a self-regulating system — though we did not understand our own place in it. I bought an annual season ticket to London Zoo and we began making daily visits to the family of orang-utans there. And saw ourselves. Our behaviour, reflected in theirs, became ape behaviour. My mind and body connected in a way it never quite had before; and the mind-body connection is even more fundamental than the art-science one. We understood we were primates. And we saw what human beings were doing to this particular small family of captive primates. I wrote about this sense of human beings as part of nature, but also separated from it, in *Light Years*.

This is a novel whose form follows the months (12 sections, 52 chapters) and tries to show the extraordinary beauty of the living world as the seasons change. But the earth is also an actor in the drama, set like a jewel among the other planets in the universe. I talk about the first human inroads into the garden of earthly delights through pollution and global warming, but *Light Years'* characterising note is hope. I did not doubt the rhythm of the seasons.

In 1998 I awarded James Lovelock a (fictional, but deserved) Nobel Prize in my futuristic novel, *The Ice People*. In 2004 I published my ninth novel, *The Flood*. It is set in an imaginary city somewhere in Europe. The grinning President has started a war on a small Arab nation, and the people of the city feel helpless to stop it. This is reflected in the rain that has been falling for months. A climactic flood is clearly on the way, but people live from day to day and ignore the fact that they wade to work knee-deep in water. It is already, perhaps, almost too late. How many more epic disasters does it take to make us stop sleep-walking on through the rising flood-water?

From time to time in *The Flood* the sun comes out, revealing moments of kaleidoscopic beauty where everything is connected to everything else. The myth of the human city as a place where only one species can live is confounded by the dance of foxes, starlings, insects. Heaven is here, in the Gaian moment. The novel asks if we must watch it disappear.

In a recent interview James Lovelock talked about the human faculty of vision. Explaining why he would regret our loss if global warming prevented our survival as a species, he said that 'It was through our eyes that [the earth] was, for the first time, able to see her beauty.' Seeing, as I do, a close relationship between the spheres of beauty, of health and of happiness, I think animals' receptivity to health and happiness probably shows they too have an appreciation of beauty, even if it differs from our own. We may not be the first or the only creatures to experience the kind of joy in feeling and seeing that I would relate to beauty.

And yet there is something important in what Lovelock said. Non-human animals are unable to record what they see and feel. That is a solely human trait. If we are something as well as accomplished destroyers, we are global thinkers (liberated by our astonishing capabilities to travel and to communicate) and recording angels. C.S. Elton and E.J. Scovell both nobly fulfilled that role: Charles counted, Joy painted.

In one of her late poems, 'The Long Grass', the poet wrote about her love for the scientist, which was based on, and expressed through, a shared love for the earth which also inspired their work.

> Our love was deep in the long grass
> As clover flowers — not deeper was.
> Not deeper are the ocean beds.
> It was all that earth needs.

To live more lightly on this planet, our mother, to try to look with her eyes, to learn from science and to teach through art: this might be a way of loving Gaia and ourselves while there is still time.

Elaine Brook

Gaia and the Sacred Feminine

Our relationship with the Earth as Mother is one of our oldest archetypes—understanding our deep connection with the living planet, and dependence on 'her' for survival. The steady decline of this worldview has brought our species to an ecological crisis point, and a dawning awareness that archetypal relationships are not merely primitive superstitions to be discarded. They are an essential part of our ability to engage in collective behaviour which furthers the common good over individual short-term priorities.

Surviving fragments of older cultures have managed to maintain their stories, lifestyle, and fabric of society despite military invasions, annexation of indigenous lands for plantations, clear-felling of forests, mining and increasing urbanisation. They offer glimpses of our own roots, and while we cannot go back to living in the past, an essential part of growing up is learning from each stage of experience and understanding. This is true for cultures as well as individuals and retracing our path to where we took a wrong turning is a prerequisite for seeing where we should be heading next, if we are to avoid ecological catastrophe for all or most species as well as ourselves.

I was privileged to spend 12 years living and working in remote areas of the Himalaya, in the company of traders, farmers, priests, and shamans. What I learned during that time enabled me to see the value of shared stories about who we are and our relationships with each other and the wider ecosystem. These stories facilitate collective and co-operative actions for the long-term benefit of society — without a sense of oppression or apparent infringements of individual rights.

The central mythic story of Himalayan culture is retold in every temple and village each year at a particular phase of the lunar calendar; a coming together of local religious teachers and the wider community. It recounts the historic encounter between ancient animist, shamanic cultures and the arrival of Buddhism. In common with other emerging religions of that era, Buddhism reflected a shift of focus to a more human-centred philosophy, in an increasingly hierarchical city-state society with those at the top no longer directly involved with the land in order to survive.

The temple dancers re-enact the arrival of tenth-century Indian Buddhist teacher Padmasambhava, and his psychic 'victory' over Tibet's old religions and priests. Instead of destroying the vanquished opposition, the victor assigned them new jobs as protectors of the new religion. As well as being an allegory for the inner journey of the spiritual practitioner, this story established relationships enabling old and new wisdom to co-exist, more or less harmoniously, up to the present day.

The shamans and the Buddhist priests each have complementary roles within their communities; the priest conducts rituals and advises on ethics, emphasising ideals of love, compassion and awareness of our interdependence with all beings. The shaman acts as intermediary between the community and wider ecosystem. Entering a trance state through meditation, dance or drumming, the shaman's consciousness aligns with the energy or flow of nature. Through rigorous training with more experienced practitioners and many days of solitary vision quests in remote and wild places, the shaman learns to understand these voices of nature and express them in language the human community can comprehend. Only then can the uninitiated collectively engage in actions to keep themselves and their ecosystem healthy and fertile for successive generations.

My own involvement with these insights made me realise the extent to which industrialised societies have moved awareness away from a whole-ecology vision to an almost entirely human-centred outlook. This shift was reinforced by increasing literacy — something our culture automatically assumes is beneficial without questioning what we may have lost in the process.

Consider the awareness invoked by scanning the series of black squiggles on the white page of a book. Reading accounts of battle, we hear the clash of swords, feel tension in danger, sadness at the slaughter of innocents, and understand how ancient cities were destroyed. Indigenous peoples without writing have that same ability to envision a breadth and depth of information and understanding about their forest from subtle signs they can see, hear, smell and touch. To them, we must seem blind and clumsy in the extreme. This is not to say that literacy is 'bad', only that human development must find ways to excel in both skills, balancing our communication both with humans and the rest of nature.

Writing itself changed, from early symbols representing elements of nature to modern forms representing human sounds. This shifts focus to the human element, and also encourages writing to be seen as something independent in itself rather than symbols linking the different topics. Perhaps it is no coincidence that the same has happened with money — something which started as an abstract notion making exchange of goods easier acquired a reified, self existing status with greater importance than the life-support system it is helping to wipe out.

Eastern mythic history is in sharp contrast to that of Western Europe, where the equivalent co-existence of newly-arrived Christianity with the older shamanic Druid culture lasted barely a hundred years before the new religion began to be used to consolidate power hierarchies. The old nature-based relationships were outlawed and demonised. Women herbalists became persecuted as witches with cauldrons and potions, and wild nature feared as the haunt of dragons.

Dragons, snakes, and other reptiles are still revered and respected in older cultures as embodiments of natural forces, fertility, and resulting wealth — a living memory that serpents had been emblems of the ancient mother goddess. Sadly, our own dragons were reduced to greedy, miserly creatures enamoured of kidnapping helpless maidens — until St George killed them off. This introduced another strand to our story, showing old ways destroyed rather than integrated into the new. The inner symbolism of the psychic struggle of positive over negative now associated 'negative' with nature.

Scilla Elworthy traces the end of the European goddess cultures:

> Soon after the first earring, the first dagger appeared … By 1250 BCE weapons and bronze tools had spread all over Europe…The big change which was to herald the end of the matrifocal societies was the notion of seizing wealth rather than creating it. This meant that every village had to build walls and learn to defend itself. Warlike tribes came south from central Asia — they were armed, fierce, and worshipped male gods. Over a period of 1000 years they gradually conquered all the agricultural, female-worshipping cultures. Possibly the last southern civilisation to fall to them was that of Crete, protected by sea on all sides …
>
> If goddess worship was to be displaced, something akin to its

own numinous power was necessary to discredit it. For example, the creation myth in Genesis reverses the idea of the female creatrix by the assertion that God (a male) made Adam (another male) and then woman, out of a rib taken from Adam's body. Then it blames the serpent, at that time known everywhere as the symbol of the goddess, for telling Eve to eat of the fruit of the tree of knowledge, another goddess symbol. So the goddess's wisdom is acknowledged, but it makes the male god so angry that he curses the serpent, and her, and the man, and puts them out of the Garden of Eden.

For many of us, there must be a moment of revelation when we suddenly recognise that the truth of religious mythologies lies not in the 'facts' they present, but arises as a reflection of the context and relationships of the time — part of an evolving drama of human consciousness learning from its collective experiences and sometimes its mistakes. For me, there was another, equally powerful insight when I realised that the end of the culture of the sacred feminine did not necessarily lead to overwhelming dominance over women and exploitation of nature as in the western industrial world. The survival of cultures which synthesise old and new stories and relationships surely holds enormous value for industrial-competitive cultures estranged from Nature and feminine.

In the context of the current ecological crisis, mass extinctions and climate change reaching tipping point, it may seem indulgent to explore ancient mythologies. Shouldn't we focus on alternative energy sources and practical economic solutions? But stories and relationships are powerful, and we cannot save what we don't love. In order to love, we need a sense and an image, and to feel this 'other' as part of our extended self.

Think of the associations conjured from the words 'witch', 'wilderness' and 'mother'. Can we learn to re-vision wilderness as 'home' and 'larder'? Or think of a woman who gathers and uses natural herbs with the same respect and confidence in her healing abilities as someone in a white coat dispensing synthetic drugs? Or extend our relationship with our human mother to our essential, life-supporting relationship with the living Earth? The unconscious mind does not relate well to analytical and abstract ideas — it needs personification in order to form strong emotional relationships. We need to find ways into relationship with Gaia in order to develop energy and inspiration for big collective changes, allowing her systems back into balance.

For this relationship to develop, mature, and remain healthy, we need to heal the division between emotion and analysis. Too long have these aspects of individual and collective psyche been kept in separate compartments, often allowing one aspect to flourish at the expense of the other. Connection and love for the planet will become painful and repressed unless expressed in practical actions — calculations to measure carbon footprints, installing insulation and solar panels, finding carbon neutral ways to travel, and so on. Technical fixes alone without the broader vision of relationship can easily degenerate into individual self-interest.

The growing popularity of accounts of ancient Mother Goddess cultures lasting millennia without warfare may facilitate the emergence of new stories. Archaeological evidence of civilised artefacts without weapons does not provide incontrovertible proof of a peaceful culture, but this may not matter. The role of mythology lies less with content but more in 'process' — the positive, synchronised behaviour pattern of the culture that espouses it.

The collective denial of industrial-consumer cultures is a natural consequence of dispensing with stories and 'personalities' that previously shaped our relationships with our human and non-human neighbours and the life-process of the living earth. Without strong relationships, we are left with a confusion of often conflicting facts and figures telling us about our life-support system. Unable to resolve the confusion we switch off and concentrate on something else rather than deal with disturbing emotions.

Denial also relates to deeper levels of stress-response for coping with crisis and trauma. When the tiger attacks, the victim's parasympathetic system switches off and sympathetic system turns on, releasing adrenaline for fight or flight. As soon as the crisis is over, this physiological change must be reversed quickly to enable sleep, digestion and healing. The memory of the trauma and loss of family members must be set aside, to be dealt with after the priorities of food, sleep and healing. This ancient survival process no longer serves us so well in the context of competitive stressful lifestyles and ongoing knowledge that we are destroying our life support system. There is no clean break for recovery and facing the trauma — it simply becomes a process of being stuck in the denial phase.

Denial leads to a mental process of 'splitting' where a genuine wish to avoid harming others is separated off during everyday actions contributing to harm. If spiritual teaching and practice is meant to enable us to be fully aware of our own mind, yet doesn't enable us to see this 'splitting' process when it happens, maybe we need to look more deeply at the way we are applying the practice.

The implications for individuals seeking a spiritual path are very real — when we meditate or pray we open the heart to interdependence and compassion. If we buy and consume in line with our industrial-growth society, we close the heart to avoid seeing the suffering we are colluding with and causing. There is a mental continuum that accompanies our everyday actions which needs to become congruent with feelings in meditation otherwise we begin to suffer from a kind of spiritual indigestion.

Older cultures, such as those in the Himalaya, are based on much smaller communities than in this age of urban sprawl and mass transport. Reciprocal relationships can form a supportive web throughout a small population. The sense of self extends to the community as a whole, beyond the individual and immediate family, supported by shared stories setting basic parameters for socially-acceptable behaviour. Acquired from infancy, these become unquestioned assumptions, while the assumptions of contemporary industrial-growth culture lead us to follow the pursuit of money and 'standard of living' as synonymous with the pursuit of happiness. Individual rights are valued above responsibilities — without making the link to the resulting problems. The Bhutanese concept of 'Gross National Happiness' instead of Gross National Product is only recently becoming better known outside Bhutan.

The advertising industry skilfully uses imagery and story-telling techniques of mythology to great effect. Unfortunately, it delivers the message that happiness comes to individuals through competition and consumerism. The idea that happiness can only be secure if shared by all, and co-operation, not competition, achieves this, has been branded 'preaching', 'worthy but dull', 'old-fashioned'.

Our written history is a form of persuasive story which mainly focuses on the politics and wars of those seeking to expand their wealth and power, and largely excluding most of the activities of the cooperative and caring. If we recognise it as 'criminal history' and rewrite the history books with emphasis on the compassionate deeds of history's true heroes, it would introduce different assumptions about human behaviour. New insights in biology show that survival of the fittest only plays a small part in evolution, mainly in the colonizing phase. Climax ecosystems become stable with a high level of co-operation and mutual benefit between organisms. This new science as collective story needs to reach more people to become effective.

It is as if many of us are suddenly waking up and realising we have been sharing the wrong dream—a dream in which we believed we were separate and independent from other living creatures and from the Earth that supports us. The difficulty for our post-mechanistic-science culture is that we cannot go back to the mythic archetypes and stories of older societies because we have been trained only to rely on empirical, tangible evidence.

Yet the Gaia story could become an inspirational theme for change. Can we find ways to re-discover our collective relationship with Gaia, Isis, Innana, Kali, Brigid, Tara? How we think of ourselves and our relationship with the biosphere will shape the delicately balanced course of humankind and our Gaian neighbours over the coming years. Scientists and artists will need to become the shamans of post-modern society, helping reconnect human communities with the voices of the Earth.

Index

SOCIETAS: essays in political and cultural criticism

Public debate has been impoverished by two competing trends
On the one hand the trivialization of the media means that in-depth commentary has given way to the ten second soundbite. On the other hand the explosion of knowledge has increased specialization, and academic discourse is no longer comprehensible.

This was not always so — especially for political debate. But in recent years the tradition of the political pamphlet has declined. However the introduction of the digital press makes it possible to re-create a more exciting age of publishing. *Societas* authors are all experts in their own field, but the essays are for a general audience. The books are available retail at the price of £8.95/$17.90 each, or on bi-monthly subscription for only £5/$10. Details at **imprint-academic.com/societas**

IMPRINT ACADEMIC, PO Box 200, Exeter, EX5 5YX, UK
Tel: (0)1392 851550   Fax: (0)1392 851178   sandra@imprint.co.uk